EXPLOSIVE ORDNANCE DISPOSAL (EOD) OPERATIONS

September 2013

DISTRIBUTION RESTRICTION. Approved for public release; distribution is unlimited.

United States Government
US Army

*ATP 4-32(ATTP 4-32)

Army Techniques Publication
No. 4-32

Headquarters
Department of the Army
Washington, DC, 30 September 2013

EOD Operations

Contents

		Page
	PREFACE	iv
	INTRODUCTION	v
Chapter 1	**EXPLOSIVE ORDNANCE DISPOSAL**	1-1
	Mission Statement	1-1
	Organization	1-2
	Army National Guard	1-7
	EOD Team Leader Certification	1-7
	EOD Staff Officer/Non-Commisioned Officer	1-8
	Joint Operational Phasing Construct	1-8
Chapter 2	**CONTINGENCY OPERATIONS**	2-1
	The Operational Environment	2-1
	Group and Battalion Operations	2-2
	Command and Support Relationships	2-3
	Joint and Multinational Operations	2-3
	Counter-IED Operations	2-4
	Company and Platoon Operations	2-4
	EOD Team Operations	2-11
Chapter 3	**DEFENSE SUPPORT OF CIVIL AUTHORITIES**	3-1
	Group and Battalion	3-1
	Company Command Section	3-2
	Company Operations Section	**Error! Bookmark not defined.**
	EOD Team Response	3-6
	Very Important Persons Protection Support Activity	3-8
	Mission Receipt	3-8
	Pre-Mission Preparation	3-9
	Mission Execution	3-9
	Post Mission Actions	3-10
Chapter 4	**SUPPORT TO SPECIAL OPERATIONS FORCES**	4-1

Distribution Restriction: Approved for public release; distribution is unlimited.

*This publication supersedes ATTP 4-32 dated 19 December 2011.

Operational Environments 4-1
75th Ranger Regiment (Airborne) 4-4
Special Forces Groups (Airborne) 4-4

Chapter 5 **CHEMICAL, BIOLOGICAL, RADIOLOGICAL, AND NUCLEAR OPERATIONS 5-1**
Chemical, Biological, Radiological and Nuclear Threats And Hazards
Response 5-1
Nuclear Accident and Incident Response and Assistance 5-8
Respond to a Radiological Attack 5-12
Respond to Depleted Uranium Incidents 5-12

Appendix A **EOD PRE-COMBAT CHECKLIST FOR MOUNTED OPERATIONS A-1**

Appendix B **EOD PRE-COMBAT CHECKLIST FOR DISMOUNTED OPERATIONS B-1**

Appendix C **SUPPORTING ORGANIZATIONS C-1**

Appendix D **CONTAMINATION/DECONTAMINATION STATION SETUP D-1**

GLOSSARY Glossary-1

REFERENCES References-1

INDEX Index-1

Figures

Figure 1-1. EOD group 1-3
Figure 1-2. EOD battalion 1-4
Figure 1-3. EOD company 1-5
Figure 1-4. EOD platoon 1-6
Figure 1-5. EOD company WMD 1-6
Figure 1-6. EOD company (CONUS Support) 1-7
Figure 1-7. Example of duty status and command relationships (ADP 3-28) 1-7
Figure 1-8. EOD support across the joint phasing model 1-9
Figure 2-1. Communication architecture 2-6
Figure 2-2. Lightweight metal detector 2-14
Figure 2-3. 0/5/25 meter check 2-17
Figure 2-4. Inner and outer cordon 2-18
Figure 2-5. Inner cordon with wedge formation 2-19
Figure 2-6. Inner cordon in urban area 2-20
Figure 2-7. Circle search pattern 2-24
Figure 2-8. Strip/lane search pattern 2-24
Figure 2-9. Zone search pattern 2-25
Figure 2-10. Grid search pattern 2-25
Figure 2-11. Scene sketch 2-26
Figure 4-1. SOF EOD Mission Space 4-2
Figure 5-1. Example downwind hazard prediction area 5-4
Figure 5-2. Example emergency personnel decontamination station (EPDS) 5-8

Figure 5-3. Example modified emergency personnel decontamination station (EPDS) 5-8
Figure 5-4. Emergency contamination control station (ECCS) 5-12

Tables

Table 1-1. EOD group within theater area of operation 1-3
Table 1-2. No EOD Group within Theater Area of Operation 1-4
Table 1-3. No EOD group or battalion within theater area of operation 1-5
Table 2-1. EOD 9 Line 2-9

Preface

This manual focuses on EOD techniques which have either been developed or changed significantly over the past 11 years. Modularity has created exponential growth of EOD forces in an extremely short amount of time. Rules of allocation for EOD companies have continually changed to adapt to the current fight and ensure that EOD forces remain relevant. This publication provides Army EOD the opportunity to capture into a single document the most successful techniques available to perform EOD tasks and effectively complete missions.

ATP 4-32 is a sole source documented explanation of EOD mission sets throughout the full range of military operations. The publication covers support provided to combatant commanders on counter terrorism and irregular warfare. EOD forces are a key enabler to deter and defeat aggression, project power despite anti-access/area denial challenges, counter weapons of mass destruction, and maintain a safe, secure and effective nuclear deterrent. EOD also supports civil and federal authorities in the defense of the homeland. Most stability, counter-insurgency, humanitarian, and disaster relief operations will require EOD support. This ATP distills public laws and regulatory guidance that specifically direct the Services EOD mission sets and provides coherent Army doctrine for EOD in unified land operations.

ATP 4-32 provides doctrinal guidance for EOD commanders and Soldiers responsible for all EOD operations. Commanders and staffs of Army headquarters serving as joint task force or multinational headquarters should also refer to applicable joint, multiservice or multinational doctrine concerning EOD operations. Trainers and educators throughout the Army will also use this manual.

Commanders, staffs, and subordinates ensure their decisions and actions comply with applicable United States, international, and in some cases, host-nation laws and regulations. Commanders at all levels ensure their Soldiers operate in accordance with the law of war and the rules of engagement. (See Field Manual [FM] 27-10).

ATP 4-32 applies to the Active Army, the Army National Guard(ARNG)/Army National Guard of the United States (ARNGUS), and the United States Army Reserve (USAR) unless otherwise stated.

The United States Army Training and Doctrine Command is the proponent for this publication. The preparing agency is the G3 Doctrine Division, USACASCOM. Send comments and recommendations on a DA Form 2028 (Recommended Changes to Publications and Blank Forms) to Commander, U.S. Army Combined Arms Support Command and Ft. Lee, ATTN: ATCL-TSD, 2221 A Avenue, Fort Lee, VA 23801, or submit an electronic DA Form 2028 by e-mail to usarmy.lee.tradoc.mbx.leee-cascom-doctrine@mail.mil

Introduction

ATP 4-32, EOD Operations, is the first EOD doctrinal publication released under Doctrine 2015. This ATP is consistent with the doctrine of unified land operations found in ADP 3-0. ADP 3-0 shifted the Army's operational concept from full spectrum operations to unified land operations.

ATP 4-32 discusses best practices and techniques developed by EOD technicians and made changes from the now obsolete 2011 ATTP 4-32. The most significant changes are the discussions on EOD support to special operations forces (SOF) and EODs role in the joint phasing model. This manual also addresses how EOD provides support throughout unified land operations. As discussed in ADRP 3-0, the doctrine of unified land operations describes how the Army demonstrates its core competencies of combined arms maneuver and wide area security through decisive action. EOD techniques used to simultaneously support decisive action are discussed throughout the manual. ATP 4-32 does not introduce, rescind or modify any new terms.

Chapter 1 provides leaders a quick tutorial on EOD roles, organization and capabilities to defeat explosive ordnance, and briefly describes EOD mission sets in planning for each phase of the combatant commander's joint phasing model. The combatant commander's respective deliberate plans contain detailed EOD planning and operational information.

Chapter 2 describes and documents the "best practices" utilization of the EOD sustainment function during crisis response and limited contingency operations. It provides unclassified lessons learned from the last 12 years of conducting EOD tasks on offensive, special operations, intelligence, protection and homeland defense and defense support to civil authorities (DSCA).

Chapter 3 explains the "how to" of EOD tasks for DSCA in support of combatant commanders and civil authorities. It emphasizes tasks that both active and reserve component EOD units perform in direct support of civilian law enforcement agencies on a daily basis.

Chapter 4 provides doctrine on EOD tasks to support Special Operations during special warfare, shared activities and surgical strike operations. It describes Army EOD's globally provided support of combatant commander response forces, special mission units, special forces groups (airborne) and the 75th Ranger Regiment (Airborne).

Chapter 5 provides doctrine on EOD operations for combating weapons of mass destruction (WMD). These highly technical EOD operations are guided by and documented in classified technical manuals with validated render safe procedures. EOD units are uniquely qualified to conduct classified counter-WMD operations as all EOD Soldiers possess top secret security clearances with critical nuclear weapon design information access.

Chapter 1
Explosive Ordnance Disposal

Explosive ordnance disposal (EOD) is a key asset in the protection of military and civilian personnel, critical assets, infrastructure, and public safety. Ordnance and explosive threats are present during all phases of joint operations. In order to manage and mitigate risk at the lowest possible level, commanders must integrate EOD during the planning and the simultaneous execution of combined arms maneuver and wide area security.

MISSION STATEMENT

1-1. Provide EOD support to unified land operations by detecting, identifying, conducting on-site evaluation, rendering safe, exploiting, and achieving final disposition of all explosive ordnance, including IED and weapons of mass destruction (WMD); provide support to joint, interagency, intergovernmental and multinational operations as required.

1-2. EOD supports the combatant commander's capability to protect the force within the following principles as outlined in Army doctrine reference publication (ADRP) 3-37, *Protection*:

- **Comprehensive.** EOD expertise provides a complete capability to identify, render safe/dispose, and exploit all facets of explosive ordnance to include US and foreign unexploded explosive ordnance (UXO), IEDs, and WMD. The adaptable force structure enables EOD to assume mission command of all like capabilities and enablers that counter and exploit the explosive ordnance and IED threat.
- **Integrated.** This crucial capability readily integrates within other protection capabilities such as providing exploitation products to support rule of law efforts, or rendering safe and supporting exploitation of WMD materials in support of CBRNE response teams. This expertise is also integrated into a combatant commander's formation by an adaptable and scalable force allocation system, providing key counter explosive ordnance enabler capabilities at all maneuver echelon levels vertically from corps to battalion and in both conventional forces and in support of special operations.
- **Layered.** Protection from all facets of UXO, IED, and WMD comes within a layered approach. EOD can provide recurring training to the general force in how to recognize and react to encountered threats and advise physical security plans how to protect critical facilities from explosive threats; EOD teams perform render safe and disposal of encountered threats; and then with support from other specialized enablers conduct collection and exploitation of recovered items and components that support intelligence and targeting processes.
- **Redundant.** The robust force structure of the EOD organization provides a redundancy of capability both vertically and horizontally within maneuver forces, with a force allocation plan of both direct support and general support EOD teams. Strategically, a significant Army National Guard EOD force structure provides a considerable reserve capability that can be focused on secondary priorities and fill requirement gaps for the active force.
- **Enduring.** The EOD force is a crucial asset within all aspects of unified land operations and within the entire cycle of joint force operations. Providing EOD expertise during military to military engagements, humanitarian demining action, and other theater security cooperation program engagements is a key component of a combatant commander's Phase 0/I Shaping and Deterrence operations. During Phase II-IV EOD provides critical protection capabilities from explosive ordnance threats and supports the identification and targeting of enemy networks thru exploitation of captured material. In the stabilization and enabling phases, security force assistance operations to build partnered nations military and civil EOD/counter-IED resources

become the priority. Additionally, EOD forces are essential to the continuing homeland defense mission and providing routine support to civil authorities; providing on and off installation support to federal, state, local public safety officials to render safe explosive ordnance/IEDs and support to the Secret Service and Diplomatic Security Service for the protection of very important persons.

ORGANIZATION

1-3. The EOD group and battalion commands and staffs may exercise synchronized mission command of Army, joint, and multinational EOD forces. Senior EOD personnel can provide the expertise to plan, prepare, execute, assess and integrate external EOD formations into the supported unit.

1-4. EOD provides the supported commander the capability to render safe and dispose of all explosive ordnance, to include UXO, IEDs and improvised explosives. The ability to render safe explosive ordnance is essential in providing intelligence to target IED networks, as well as update TTPs, both enemy and friendly. EOD personnel have the ability to dispose of all types of ordnance in the safest manner possible. Ordnance that is disposed of improperly can cause damage to both military and civilian personnel and property within the blast and fragmentation zone.

1-5. EOD commanders and staffs from group to platoon provide the supported commander and staff not only mission command of EOD subordinate units, but also other specialized enablers such as technical intelligence and exploitation assets focused on weapons and ordnance. They also provide a special staff capability and subject matter expertise concerning all facets of explosive ordnance, IEDs, CBRN/WMD threats and hazards, and technical exploitation. The EOD leadership and staff can support or provide staff lead of functional working groups (e.g. counter-IED), integrated into supported unit staff planning processes, provide support and input into targeting and the intelligence process, and lead exploitation activities and processes.

EOD Group

1-6. The EOD group is a functional mission command headquarters for EOD operations. The group conducts staff planning and technical control of all EOD assets in a theater and provides EOD staff liaison to the ASCC. The EOD group is capable of conducting EOD mission command for two to six EOD battalions. The EOD group is attached or placed OPCON, to coordinate counter-IED and weapons technical intelligence operations, to a theater army, corps, or JTF in support of a specific operation, operation order, operation plan, or concept plan. The group may also form the core of a specialized combined JTF with mission of various protection and exploitation enablers such as counter-IED, exploitation, or counter-WMD task forces. The group can also provide enabling support, analysis, and advisement to execute targeting efforts, theater exploitation, and counter-WMD in order to provide maneuver support and force protection in all operational environments.

Explosive Ordnance Disposal

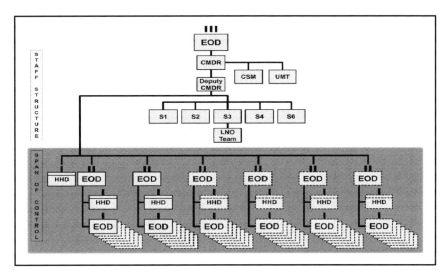

Figure 1-1. EOD group

Table 1-1. EOD group within theater area of operation

EOD Organization	Modeling Rule of Allocation	Supported Organization	Relationship
⩥EOD⩤ (III)	1 per Theater Army 1 per Corps 1 per Joint Task Force 1 per Combined Joint Task Force 1 per Homeland Defense 1 per 2-6 EOD Battalions	Theater Army Corps Joint Task Force Combined Joint Task Force	Attached/OPCON Attached/OPCON OPCON OPCON
⩥EOD⩤ (II)	1 per Division 1 per Joint Task Force 1 per Combined Joint Task Force 2 per Homeland Defense 1 per 3-7 EOD Companies	EOD Group Division Joint Task Force Combined Joint Task Force	OPCON DS/GS DS/GS DS/GS
EOD (I)	1 per BCT 1 per SFG(A) 1 per Ranger Regiment 8 per Homeland Defense 1 per 1-5 EOD Platoons	EOD Battalion BCT MEB SFG(A) Ranger Regiment	OPCON DS/GS DS/GS OPCON/TACON OPCON/TACON
••• EOD	3 Per committed BCT 1 per Special Forces Battalion 1 per Ranger	EOD Company Maneuver Battalion Special Forces Battalion Ranger Battalion	Assigned DS/GS DS/GS DS/GS

Chapter 1

Table 1-1. EOD group within theater area of operation

EOD Organization	Modeling Rule of Allocation	Supported Organization	Relationship
	Battalion 24 per Homeland Defense 1 per 3 EOD Teams		

EOD BATTALION

1-7. The EOD battalion is a functional mission command headquarters for EOD operations. The EOD battalion conducts staff planning, staff control of all counter-IED assets within a division area of operations. The EOD battalion is capable of conducting EOD mission command and supervision of EOD operations for two to seven EOD companies. The EOD battalion may be attached or OPCON to a theater army, corps, division, JTF, or CJTF in support of a specific operation, operations order, operation plan, or concept plan.

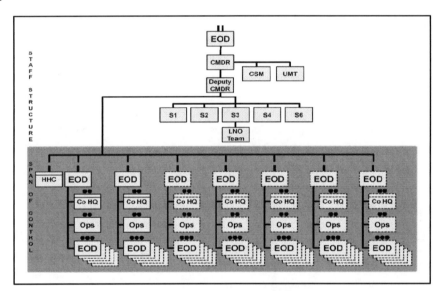

Figure 1-2. EOD battalion

Table 1-2. No EOD Group within Theater Area of Operation

EOD Organization	Modeling Rule of Allocation	Supported Organization	Relationship

EOD (battalion)	1 per Division 1 per Joint Task Force 1 per Combined Joint Task Force 2 per Homeland Defense 1 per 3-7 EOD Companies	Theater Army Corps Division Joint Task Force Combined Joint Task Force	Attached/OPCON OPCON OPCON OPCON OPCON
EOD (company)	1 per BCT 1 per SFG(A) 1 per Ranger Regiment 8 per Homeland Defense 1 per 1-5 EOD Platoons	EOD Battalion BCT MEB SFG(A) Ranger Regiment	OPCON DS/GS DS/GS OPCON/TACON OPCON/TACON
EOD (platoon)	3 Per committed BCT 1 per Special Forces Battalion 1 per Ranger Battalion 24 per Homeland Defense 1 per 3 EOD Teams	EOD Company Maneuver Battalion Special Forces Battalion Ranger Battalion	Assigned DS/GS DS/GS DS/GS

COMMAND AND SUPPORT RELATIONSHIP

1-8. The EOD group or battalion may deploy as the senior mission command element for EOD operations and may be organized under a theater army, corps, division, JTF or CJTF. The EOD battalion reports directly to the EOD group and may deploy with its assigned EOD group or may have OPCON designated to a separate EOD group.

EOD COMPANY

1-9. The EOD company provides mission command of one to five EOD platoons and provides administrative company level planning and support based on the level of employment to include augmenting the brigade combat team (BCT) commanders with a special staff element. The EOD company provides EOD service throughout the theater AO and direct support to designated BCT/SFG(A).

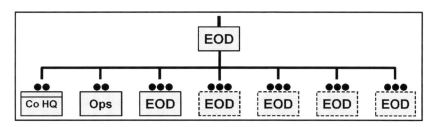

Figure 1-3. EOD company

Table 1-3. No EOD group or battalion within theater area of operation

EOD Organization	Modeling Rule of Allocation	Supported Organization	Relationship

Chapter 1

EOD (I)	1 per BCT 1 per SFG(A) 1 per Ranger Regiment 8 per Homeland Defense 1 per 1-5 EOD Platoons	BCT SFG(A) Ranger Regiment	OPCON OPCON/TACON OPCON/TACON
EOD (•••)	3 Per committed BCT 1 per Special Forces Battalion 1 per Ranger Battalion 24 per Homeland Defense 1 per 3 EOD Teams	EOD Company Maneuver Battalion Special Forces Battalion Ranger Battalion	Assigned DS/GS DS/GS DS/GS

EOD PLATOON

1-10. The EOD platoon is normally employed at the battalion level and provides leadership, supervision and technical guidance for three to four EOD teams, typically consisting of three personnel. The EOD platoon provides the capability to eliminate and reduce explosive, chemical, biological, and nuclear hazards, including IEDs and conventional U.S. and foreign UXO. The platoon provides support to the US Secret Service and Department of State in protection of the President, Vice President and other dignitaries as directed.

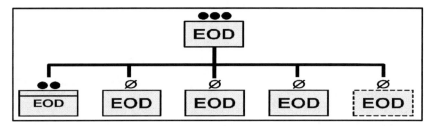

Figure 1-4. EOD platoon

EOD COMPANY WMD

1-11. The EOD WMD company provides highly technical EOD operations and containment procedures for WMD in support of joint or interagency operations. It has the ability to respond anywhere in the world with two fully capable eight person platoons as part of the Joint Technical Operations Team to defeat or mitigate the effects of WMD against the US. The unit has the capability to provide four WMD platoons to support the Army or other US agency in support of missions to defeat or mitigate WMD directed against the US or national interest.

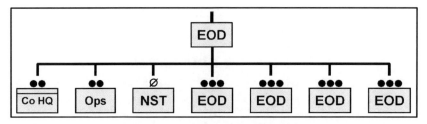

Figure 1-5. EOD company WMD

EOD COMPANY (CONUS SUPPORT)

1-12. The EOD company (CONUS Support) provides EOD service in the reduction and elimination of hazardous munitions and explosive device services to federal, state, and local agencies on an area basis. The EOD (CONUS support) Company is allocated based on the concept of support requirements. Dependent on the ordnance battalion EOD for administrative, CBRN, religious, legal, personnel, field feeding, and supply services/support, the subordinate elements of the support maintenance company to perform field maintenance on the unit's organic equipment and the area support medical company for force health protection and Class VIII support.

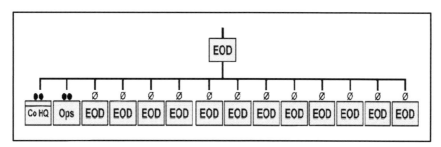

Figure 1-6. EOD company (CONUS Support)

ARMY NATIONAL GUARD

1-13. Administered by the National Guard Bureau (a joint bureau of the departments of the Army and Air Force), the Army National Guard (ARNG) has both a federal and state mission. The dual mission, a provision of the US Constitution and the US Code of laws, results in each Soldier holding membership in both the National Guard of his or her state and in the US Army. ARNG EOD assets are resident within several states and the Commonwealth of Puerto Rico.

1-14. In addition to ARNG deployments in support of federal missions, the ARNG plays an extensive and highly visible domestic role. As part of its unique "dual-mission" responsibilities, the ARNG routinely responds to domestic requirements within each state. Whenever disaster strikes or threatens, the ARNG represents the most significant asset governors can rapidly mobilize to provide protection, relief, and recovery. In accordance with the applicable laws, Department of Defense Policies and Department of Defense Directives, the ARNG may provide support of special events and defense support and assistance to civilian law enforcement agencies. Figure 1-7 illustrates the duty status and command relationships of regular Army forces and National Guard.

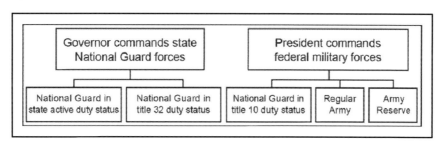

Figure 1-7. Example of duty status and command relationships (ADP 3-28)

EOD TEAM LEADER CERTIFICATION

1-15. EOD Soldiers go through a certification process before becoming team leaders. If an EOD Soldier has been certified as a team leader they are given the ability to make judgment calls based on the OE and threat. A team leader has to make many difficult decisions throughout the course of a combat tour. Rendering safe and disposing of explosive ordnance, IEDs and homemade explosives in any environment can cause catastrophic consequences. Unless their actions are proven to have been negligent, a team leader must feel confident that their decisions will be supported by their leadership even in circumstances that may not have the desired outcome.

1-16. The team leader is the most coveted position for an EOD technician. There are few jobs that give such a large amount of responsibility to a Soldier. The team leader has the responsibility to mentor and develop their team members. The team leader must teach their team members how to perform all of the responsibilities as a team leader. This will not only allow them to prepare to be team leaders but will allow them to be able to take control of a scene in the case that the team leader becomes a casualty.

EOD STAFF OFFICER/NON-COMMISIONED OFFICER

1-17. The EOD staff officer/NCO is a key link between the corps/division commander and the EOD forces that integrate into the maneuver fight. In order to perform these duties the officer/NCO must have a keen understanding of joint/Army EOD doctrine and be able to articulate the capabilities, constraints and limitations of an EOD unit.

1-18. The EOD staff officer/NCO should focus on EOD capability and the resourcing, integration and reallocation of the EOD capability into the corps/division commander's AO. They will also focus on the render safe procedure (RSP)/technical exploitation expertise and the application of that expertise.

1-19. The EOD staff officer/NCO is best utilized in the G-3 section as the EOD/counter-IED subject matter expert. The position should focus on integration and understanding the corps/division staff, different staff functions, predeployment training and integration, and work the integration of counter-IED related training. The position will also assist the G-3 throughout the military decision making process in order to make well informed recommendations to the commander.

1-20. During the corps/division subordinate BCT road to war training, the EOD staff officer/NCO may advise on all counter-IED specific training to be conducted in accordance with not only US Army Forces Command regulations, but theater requirements as well.

1-21. The personnel assigned to this position need to maintain contact with the closest EOD battalion or group in order to keep EOD leadership informed of all current and future operations within the corps/division. While deployed, the EOD staff officer/NCO needs to maintain relationships with whoever generates EOD capability. This will keep the corps/division commander informed of all incoming, on the ground, and outgoing EOD units, from all services, operating within the corps/division AO.

1-22. While deployed, it is recommended that this position remain in the G-3. The EOD officer will be the lead for the counter-IED working group (ADRP 3-37). If there is not a counter-IED working group in the corps/division protection cell it is recommended to integrate with protection working group, the anti-terrorism working group, or the CBRN working group.

JOINT OPERATIONAL PHASING CONSTRUCT

1-23. The phased construct is utilized in campaign planning to provide intermediate goals in order to accomplish the overall objective of a successful campaign (Joint Publication (JP) 5-0). Phases are distinct in time, space, and/or purpose from each other and represent a natural progression. Though the phases are designed to be conducted sequentially, EOD mission sets, like other activities, may begin in a previous phase and continue into subsequent phases.

Explosive Ordnance Disposal

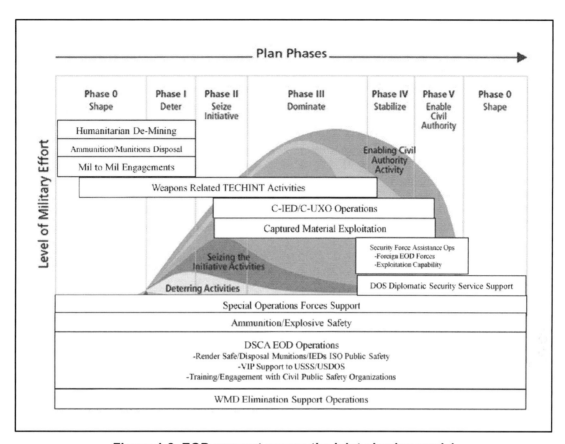

Figure 1-8. EOD support across the joint phasing model

PHASE 0 SHAPE

1-24. Joint and multinational operations, including normal and routine military and interagency activities, which are performed to dissuade or deter potential adversaries and to assure or solidify relationships with allies. EOD mission sets that support Phase O operations include the following.

Humanitarian De-Mining Support

1-25. The Department of Defense (DOD) Humanitarian Mine Action (HMA) Training Program is authorized by Title 10 US Code-Section 407 and is planned and executed by geographic combatant commanders (GCC) as part of the theater security cooperation plan. The Director, Defense Security Cooperation Agency manages HMA and is responsible for the execution of the overseas humanitarian, disaster, and civic aid appropriation. The Office of the Secretary of Defense provides policy oversight and coordinates with Department of State (DOS) weapons removal and abatement program.

1-26. EOD has been identified as the "Priority Force" of choice according to CJCSI 3207.01B as they lend themselves best for conducting "Train-the-Trainer" type HMA missions. Their expertise in all types of UXO and their routine requirement to train and teach general purpose forces match perfectly with the HMA mission set.

Military to Military Engagements

1-27. Traditional combatant commander activities such as military to military engagements are events that EOD units will continue to provide invaluable contributions. Such events by Army EOD units allow for assessments of a partnered nation's capabilities in counter-IED or against other explosive ordnance threats, ammunition storage procedures and facilities, as well as their EOD organizations. Such engagements will be key tasks within all echelons from platoon to group headquarters.

1-28. Teams will be utilized to conduct training and assessments of partner nation's capabilities, such as their ability to conduct exploitation, post blast analysis, render safe, or ammunition handling techniques. EOD battalion and group staffs will be used to assess and teach how to establish and operate exploitation capabilities, conduct analysis and other facets of the exploitation process, and feed the intelligence cycle and the targeting process. These assessments allow for follow on training and partnering events to increase or sustain the nations capabilities which strengthen overall operational and strategic relationships.

Support to Foreign Internal Defense (FID) and Security Force Assistance (SFA) Activities

1-29. EOD expertise will be increasingly important to SOF efforts to build and sustain a friendly nation's ability to maintain a free and protected society from subversion, lawlessness, insurgency, terrorism, and other threats to its security and to US general purpose forces ability to support the development of the capacity and capability of foreign security forces and their supporting institutions. According to the US Army Counter-IED Strategy the IED will remain an effective casualty producing device for the foreseeable future, therefore it is likely that EOD support to these operations will increase. The skill set best suited to train a country's force to counter these threats is inherent within the EOD force structure.

JPAC Support

1-30. Joint Prisoner of War (POW)/Mission in Action (MIA) Accounting Command (JPAC) conducts global search, recovery, and laboratory operations to identify unaccounted-for Americans from past conflicts in order to support the DODs personnel accounting efforts. EOD personnel provide a crucial skill set to this high visibility mission by ensuring the safety of recovery team members and the local populace by eliminating encountered explosive ordnance, either from the payload of downed aircraft or residual UXO found on site.

PHASE I. DETER

1-31. Demonstration of the capabilities and resolve of the joint force to deter undesirable adversary action; includes activities to prepare forces/set conditions for deployment and employment if deterrence is not successful. EOD mission sets that support Phase I operations include the following.

Habitual Training Relationships with conventional and special operations forces

1-32. In addition to the previous operations, EOD forces will support Army activities within this phase by fostering habitual training relationships with their respective direct support maneuver echelon by participating in maneuver unit training cycle to include local installation training and combat training center rotations. This participation will also include larger scale national/international exercises, command post exercises, and other events. This phase includes operational and strategic planning efforts within all levels of command from brigade to ASCC; efforts that will require the appropriate expertise to ensure key EOD capabilities are included within operational and strategic planning.

Technical Intelligence Activities

1-33. EOD personnel resident in several intelligence organizations will continue to provide requisite expertise to continual collection efforts of adversarial weapons and ordnance within the weapons technical intelligence process. These activities are necessary to maintain the advantage over current or possible adversarial weapons and technology. EOD expertise is necessary to developing counter measures and to producing a clearer intelligence picture of threat weapons and capabilities. EOD teams are the primary collectors of information on explosive ordnance and perform the first technical assessment of found or captured weapons and ordnance. Over the last 10 years the EOD group has been the primary mission command element for all weapons related activities as a subordinate task force.

Counter WMD Support Operations

1-34. EOD forces are key contributors within several military mission areas of the Army's capability to conduct counter WMD operations. As previously discussed, support to security cooperation, cooperative threat reduction and partner activities is a continuing strength of EOD units. EOD provides critical subject matter expertise to conduct engagements with international counter parts to improve or promote defense relationships and capacity of multinational and partner nations to execute counter WMD operations.

1-35. The other military mission area that EOD is a key component is WMD elimination. EOD provides significant expertise within all stages of elimination, to include, diagnostic and disablement, supporting technical intelligence collection and exploitation, providing guidance on protective measures, conducting render safe, assisting in destruction of transfer activities, and supporting monitoring and redirection efforts. These efforts support general purpose forces and Tier 1.

PHASE II/III. SEIZE INITIATIVE/DOMINATE

1-36. Application of appropriate force capabilities to seize the initiative; Control of the OE or breaking the enemy's will for organized resistance. EOD mission sets that support Phase II/III operations include the following:

EOD Render Safe/Disposal of Explosive Ordnance, UXO, IEDs, Homemade Explosives, CBRN Ordnance

1-37. EOD core skills are essential to the maneuver commander to protect the force, civil population, and critical infrastructure from the wide range of explosive ordnance to include UXO, IEDs, and CBRN hazards. EOD teams and platoons will provide the render safe and disposal of the hazard to ensure the mobility and protection of maneuver forces.

1-38. Companies through battalion and group will provide mission command of multiple units to ensure this capability is positioned to support maneuver commander priorities. Depending on the breadth of maneuver forces employed, EOD forces will provide the necessary mission command structure to manage the counter explosive ordnance/IED fight and to provide operational and intelligence insight into threat weapons and tactics.

1-39. Group headquarters to support corps/divisions; battalion headquarters to support divisions/task forces; companies to support brigades/BCTs; platoons to maneuver battalions is the general construct of support but can be modified to meet the maneuver commander's requirement and to counter the level of threat.

Captured and Recovered Ammunition/Munitions

1-40. EOD units have a major role in the collection, assessment and disposal of captured and recovered ammunition/munitions. This includes captured, abandoned or recovered ammunition, supply points, field storage and licensed depots, factories and testing facilities. In the absence of Ordnance (Ammunition) units, EOD units can take charge of these efforts. Refer to ATTP 4-32.2, UXO Multi-Service Tactics, Techniques, and Procedures (MTTP) for Unexploded Ordnance.

1-41. Recovered munitions are typically complete devices containing explosives, propellants, pyrotechnics, initiating composition, or nuclear, biological or chemical material for use in military operations, including demolitions which when located in the OE, in filed storage sites and licensed storage areas have not been primed for use and may or may not be in their primary or logistic packaging.

Removal of Stuck Rounds and Downloading of Misfired Munitions

1-42. Removal of stuck rounds in mortars, artillery tubes, and other weapon systems is a routine EOD operation. Prior to requesting EOD assistance for stuck rounds, the using unit must first perform misfire procedures stated in the weapon's technical manuals.

1-43. EOD can assist in downloading misfired munitions from aircraft and armored vehicles weapons systems, both foreign and US. If an enemy aircraft or armored vehicle is captured, the capturing unit should immediately request EOD assistance in order to remove all explosive hazards.

Management of Special Programs

1-44. The EOD group or battalion manages all weapons or ordnance related special programs within an AO. EOD personnel may assist commanders with any concerns regarding special programs.

Support to Special Operations (SOF Core Operations/Activities)

1-45. In addition to support to SOF, specialized EOD organizations provide direct support to other special operations forces in the conduct of all the SOF core operations and activities. The proliferation of the IED and other explosive ordnance threats drives the emerging requirement for the continued support to all SOF units. These type operations not only provide a required capability to the SOF organizations supported, they also provide other unique skill sets, experience, and leadership development that benefits the overall EOD force.

PHASE IV. STABILIZE

1-46. Performance of limited local governance, integrating other supporting/contributing multinational, intergovernmental organization, non-governmental organization, or US government agency participants until legitimate local entities is functioning. EOD mission sets that support phase IV operations include the following.

Perform Battlefield Clearance

1-47. At the conclusion of phase III the OE may contain large quantities of UXO, mines, captured enemy ammunition/munitions, homemade explosives and production labs, battle damaged military vehicles, abandoned ammunition/munitions and foreign stockpiles. EOD units are the only personnel in the US government qualified to identify and take action on foreign ordnance. EOD units will assist in re-building partner EOD capability; coordinate with DOS Office of Weapons Removal and Abatement and non-governmental organizations to establish indigenous HMA, munitions control and EOD capabilities.

Captured Material Exploitation

1-48. The most significant emerging requirement and capability that has developed from the past decade of conflict is the exploitation of captured enemy material, to include IED components. The harvesting of key forensically valuable information from material handled by enemy combatants has dramatically affected the capability of US forces to identify and target key enemy actors and networks.

1-49. EOD forces have been on the forefront and often times the lead on most aspects of this effort. The EOD community's ability to use intelligence developed from the very weapon that the enemy employs against US forces and turn it against them is vital to any counter insurgency or other stabilization operation. Though this has been primarily important during stabilization activities, this capability will also be crucial to other operations in previous phases in order to better identify and target the enemy network.

1-50. This capability includes the ability of the EOD team to start the exploitation process by conducting site exploitation. EOD expertise resident within exploitation labs and commanders and staff at all levels of EOD formations continue the exploitation, weapons technical intelligence analysis and dissemination of results to the supported maneuver staff to aid in the targeting of enemy networks.

Mission Command of Functional Task Forces (Counter-IED, WMD-E, Asymmetric Threats)

1-51. The other considerable capability emerging from Iraq and Afghanistan is the requirement to provide mission command for functional task forces, at both the battalion and group headquarters levels. Though it is unknown if the Army will conduct operations in the future in the same manner as was conducted in Iraq and Afghanistan, it is imperative that the lessons learned from conducting CJTFs TROY and PALADIN are captured and incorporated into current EOD organizational capabilities.

1-52. EOD group and battalion headquarters are uniquely suited with structure and expertise to provide the core architecture to build a task force around, as well as accept bolt on capabilities to enable the employment of an organization focused on protecting the force from explosive threats and enabling the effective targeting of the enemy using them.

1-53. In addition to controlling the subordinate EOD organizations, the functional task force commander will also mission command other specialized enablers that could include exploitation laboratories, additional exploitation teams, operations research/systems analysts, and focused intelligence analysts.

1-54. The combination of EOD expertise with other specialized enablers allows the task force to conduct focused threat (such as IED) network analysis and targeting, provide direct support packages of specialized enablers to maneuver commanders and staff, and organize and conduct unit level explosive threat focused training programs.

Department of State Diplomatic Security Service Support

1-55. EOD may be tasked with providing dedicated support to the Department of State Diplomatic Security Service in order to protect the US embassy and its personnel. This tasking will be at company level with the regional security officer obtaining DS of the designated company. The GCC will retain administrative control of the EOD company.

1-56. The mission of the EOD company will be to identify possible explosive threats and render safe any actual devices. EOD teams may also be tasked to provide support to local nationals or other embassies depending on the Department of State mission posture. This includes training, partnership events, and EOD technical support for explosive devices found off Department of State property.

1-57. EOD personnel will operate under administrative and technical diplomatic status, which does not confer full "diplomatic immunity" excusing serious crimes, but instead confers the same limited legal protection that all mission personnel have.

PHASE V. ENABLE CIVIL AUTHORITY

1-58. Joint force support to legitimate civil governance in theater; enable the viability of the civil authority and its provision of essential services to the majority of people in the region. EOD units help build indigenous military and police EOD and exploitation capability. EOD support to Phase V is a continuation of missions performed in previous phases.

Chapter 2
Contingency Operations

During contingency operations, EOD Soldiers are critical enablers that bring a specialized set of knowledge, skills, and abilities to an AO. Working in teams of two to three technicians, an EOD team provides the supported commander with the capability to render safe and dispose of all types of explosive ordnance, IEDs and homemade explosives. EOD also provides technical information that contributes to the targeting process and allows for a rapid change in TTPs based on enemy activity. Post blast investigations conducted to recover blast remnants and evidence enables commanders to gain a better understanding of the threats that their units are encountering.

THE OPERATIONAL ENVIRONMENT

2-1. The OE is a composite of the conditions, circumstances, and influences that affect the employment of capabilities and bear on the decisions of the commander (JP 3-0). The EOD Soldier must have a keen understanding of the OE and be able to make quick decisions based on the conditions, circumstances, and influences around them. The EOD team supporting operations needs to recognize the implications, and consider the long term effects that their actions can cause. Depending on the phase of the operation, the EOD Soldier will have to determine what level of risk is acceptable in order to complete the mission.

2-2. To operate effectively under conditions of uncertainty and complexity in an era of persistent conflict, future EOD forces and leaders must strive to reduce uncertainty through understanding the situation in depth, developing the situation through action, fighting for information, and reassessing the situation to keep pace with the dynamic nature of conflict. Accomplishing challenging missions and responding to a broad range of adaptive weapons and explosive threats under conditions of uncertainty will require EOD forces that exhibit a high degree of operational adaptability. EOD forces must also hone their ability to integrate joint and interagency assets, develop the situation through action, and adjust rapidly to changing situations to achieve operational adaptability.

2-3. During contingency operations, EOD personnel will find themselves in several OE throughout the course of a day. This can include having to render safe an IED in an urban area, where there is a high risk of small arms fire, or clearing and disposing of a weapons cache in an unpopulated area. It is essential that the EOD team leader understand the supported commanders objectives in order to make the appropriate decision when conducting EOD operations based on the OE that they are in.

MISSION VARIABLES

2-4. Every EOD response is a separate mission. These missions include deliberately planned missions that may last for several days or weeks, or emergency response missions that may last as little as an hour. An EOD team may go on several missions a day. Each mission has different considerations; Mission, enemy, terrain and weather, troops and support available, time available and civil considerations (METT-TC) are the mission variables that the EOD team uses when assessing what actions are appropriate.

2-5. These variables will determine the types of tools, actions on scene, and support needed in order to accomplish the mission. An EOD team may be able to use a robot to perform remote procedures on an explosive device during a response and then have to dismount and move to an area with only a select amount of tools on their next response. Because of the nature of EOD operations, teams must always be prepared to respond to several different situations in different environments.

GROUP AND BATTALION OPERATIONS

2-6. EOD groups exercise mission command of EOD battalions and companies as well as counter-IED assets and operations in an AO. EOD battalions execute mission command for lower levels of allocation, or during late phase stabilization operations, based on volume of missions and span of control. Attack the network, defeat the device and adapt the force are secondary concerns outside of protection.

2-7. The EOD group and battalion commands and staffs may exercise synchronized mission command of Army, joint, and multinational EOD forces. EOD planners can provide the expertise to plan, prepare, execute, assess and integrate external EOD formations into the supported unit.

2-8. The EOD group and battalion may form the backbone of a special purpose JTF. The JTF enables unity of effort and integrates functional capabilities and capacity throughout a theater area of operations to include; counter-IED, UXO, and WMD. This is accomplished by providing problem focused expertise, delivering relevant and ready specialized capabilities to commanders, and ensuring integration and mission execution by those capabilities. The JTF, if assigned, could articulates and prioritize explosive ordnance requirements; integrate DOD and multinational forces to build enabler capabilities for the commander; develop and enforce plans, policies, explosive safety and planning standards; and collect and disseminate lessons learned to the generating force and DOD strategic leadership.

2-9. The special purpose JTF creates fusion among organizations and enablers to include interagency partners and leverages core competencies toward defeating explosive ordnance/IED/homemade explosive/WMD as a weapon of strategic influence by integrating the following capabilities:

- Synchronizes and integrates theater level counter-IED organizations to include:
    - Weapons intelligence teams.
    - Combined Explosives Exploitation Cell.
    - Counter Insurgency Targeting Program.
    - Counter-IED Operations Intelligence Integration Center.
    - Counter-IED training teams.
    - Technical escort detachment.
    - Operational research/system analysis.
    - Counter radio controlled IED electronic warfare (CREW).
    - Law enforcement professionals.
    - Joint Expeditionary Team.
    - Explosive Hazard Coordination Cell.
    - Host nation EOD partnership.
    - Mine action center.
    - Exploitation cell.
    - Science and technology.
- Tracks critical counter-IED assets, monitors and recommends changes in priorities.
- Manages theater level IED elimination operations.
- Manages counter-IED special programs.
- Performs theater level counter-IED planning, integrating, coordinating, analysis and tasking functions.
- Manages, integrates, coordinates and synchronizes theater level exploitation functions in regards to IEDs (detection, collection, processing, exploitation, analysis and dissemination).
- Synchronizes efforts and analyzes trends among counter-IED operations, captured enemy ammunition/munitions operations, recovered munitions operations, and UXO operations.
- Provides oversight of all Joint EOD operations in a theater.
- Tracks and manages all Joint EOD assets in a theater.
- Performs theater level EOD planning integrating, coordinating and tasking functions.
- Facilitate and/or lead counter-IED and protection working groups with supported maneuver elements/staff.

- Conduct specialist (EOD/IED) in theater and partnering training.

Recent examples:
- CJTF Troy exercises command and control of specialized IED defeat forces, as well as coordinates corps level IED defeat operations, intelligence, technology, and training initiatives throughout the Iraqi theater of operations to defeat the IED system.
- CJTF Paladin fuses IED defeat related, intelligence, effects, training, equipment, and EOD forces into one organization, assigned to CJTF-76, throughout the Afghanistan theater of operations to defeat the IED system.

COMMAND AND SUPPORT RELATIONSHIPS

2-10. The EOD group provides mission command for all Army EOD assets and operations in theater. The EOD battalions provide mission command, mission tasking, technical intelligence collection and management, and limited administrative and logistic support for up to seven EOD companies. EOD battalions may deploy as the senior mission command element for Army EOD operations.

2-11. EOD companies remain under the command of their parent battalion, and/or the senior EOD mission command element in theater. Depending on the operational situation, the EOD company may be placed under the tactical control/operational control (TACON/OPCON) of another unit. When using the TACON/OPCON command option, the parent battalion retains administrative control (ADCON) of their subordinate companies. EOD companies provide general support-reinforcing on an area basis or direct support to specified elements in support of operations. Responsibilities of the EOD commander at all levels include:

- Recommending policy and distributing EOD assets.
- Monitoring EOD support missions and establishing workload priorities.
- Serving as point of contact for technical intelligence coordination.
- Coordinating general support-reinforcing and direct support.
- Ensuring each EOD unit establishes provisions for communications at each level to support EOD operations.
- Supplementing other theater force-protection procedures to meet the existing threat.
- Coordinating administrative and logistical support, as required, from the supported command.

JOINT AND MULTINATIONAL OPERATIONS

2-12. Service components will task organize and deploy with their own EOD assets. In many situations, the combatant commander, through their directive authority for logistics, can achieve economy of effort by organizing their EOD forces using common servicing. Common servicing may allow the commander to provide more efficient and effective EOD support to the joint force depending on the operational scenario. The commander should also include integration of multinational, host nation EOD forces, and contracted UXO disposal assets in a joint/multinational EOD task force.

2-13. Multinational EOD forces are not always trained to the same standards, or equipped with the same technology, of US EOD forces. Before employing multinational EOD forces, the commander must ensure that their capabilities and limitations are understood by all units that may work in an AO that is covered by multinational EOD forces. Standards for multinational EOD forces can be found in Standardization Agreement (STANAG) 2389, Minimum Standards of Proficiency for Trained Explosive Ordnance Disposal Personnel.

2-14. The United States Army Corps of Engineers, Military Munitions Division provides technical expertise for all aspects of the munitions response process. One of its missions is to support the US Army's captured enemy ammunition program. US Army Corps of Engineers has oversight of all UXO disposal contracted support. The contracting officer or contracting officer's representative must ensure commercial UXO disposal firms meet an acceptable level of training and equipment standards as determined by the Department of Defense Explosive Safety Board (DDESB) and US military EOD experts.

Chapter 2

2-15. For more information on the employment of joint and multinational EOD forces refer to ATTP 4-32.16, MTTP for EOD and North Atlantic Treaty Organization Allied Technical Publication 72, Inter-Service Explosive Ordnance Disposal Operations on Multinational Deployments.

COUNTER-IED OPERATIONS

2-16. Designing a successful counter-IED operation is a complex task that involves all echelons of the joint force and is based on a framework designed to assure the freedom of movement of friendly forces and enable commanders and staffs to plan and take proactive measures to identify and defeat IED events before they are successfully employed. Within each level of war there are key IED activities that influence operational planning. These activities must be viewed both individually and in the context of their relationship to the other activities that enable a counter-IED effort.

2-17. Counter-IED operations must take a holistic approach that incorporates intelligence, information, training, operations, materiel, technology, policy, and resourcing solutions. This approach is designed to address all of the fundamentals of assured mobility, to include prediction, detection, prevention, neutralization, and mitigation. To a larger extent this approach should be considered in terms of joint interdiction, which encompasses assured mobility as well as many other factors of warfare. Counter-IED operations are conducted across the phases of a military operation (shape, deter, seize the initiative, dominate, stabilize, and enable civil authority) and should be executed within multiple lines of operation, each to commence conditionally, and then to continue in parallel throughout a campaign. The counter-IED commander may change the lines of operation based on the OE. The three baseline counter-IED lines of effort are as follows

ATTACK THE NETWORK

2-18. Joint force attack the network actions prevent the emplacement of the IED by attacking adversary vulnerabilities at multiple points. Key vulnerabilities within the adversaries IED employment system include: abilities to influence the support of the local populace, employment of IED TTP; the ability to maintain an IED component supply and distribution chain; the ability to establish and modify IED build-emplacement process.

DEFEAT THE DEVICE

2-19. The goal when defeating an IED is to prevent or mitigate its physical effects while marginalizing or preventing the adversary from exploiting its psychological effects through information activities, such as propaganda and disinformation. Counter-IED defeat actions begin once the device has been emplaced and include detection, disarming it safely, recording the technical categorization and tactical characterization so thorough technical and forensic exploitation and analysis is achieved.

ADAPT THE FORCE

2-20. Commanders must ensure the force is adequately trained prior to deployment. Areas of special interest include the development of relevant and current IED related TTP, drills, and standard operating procedures (SOP). Training should be designed to enhance individual and unit protection and the unit's ability to effectively operate in a high-threat IED environment. Training should also include those activities that facilitate the establishment and growth of multinational and partner nation IED defeat capabilities, including the transfer of counter-IED technology and US force TTP.

2-21. For more information on counter-IED operations and IED Defeat refer to JP 3-15.1, Counter-Improvised Explosive Device Operations; AJP 3-15, Allied Joint Doctrine for Counter Improvised Explosive Devices and Training Circular 3-90.119, US Army Improvised Explosive Device Defeat Training.

COMPANY AND PLATOON OPERATIONS

2-22. The EOD company headquarters provides mission command to the platoons and teams. In the current OE, the decentralized execution of mission orders is handed all the way down to the EOD team. It is crucial that every EOD Soldier understand both the supported commander's intent as well as the EOD company commander's intent, in the case that they have to take control of an incident site.

2-23. The EOD company leadership should maintain a physical presence in the supported brigade's command post, in order to maintain continuous integration into brigade planning and operations. EOD requires both a non-classified and secret internet protocol router network in their command post area. The company maintains 24 hour communication with the brigade command post in order to maintain situational awareness of all operations that EOD is involved in.

2-24. When first assigned to provide direct support to a maneuver unit, the company commander should brief the entire spectrum of EOD operations to the brigade command team, staff and subordinate commanders. This is to ensure effective utilization of the EOD platoons and teams and make the staff aware of all capabilities and limitations of the EOD platoon. The briefings should address:

- IED defeat and exploitation.
- UXO render safe/disposal.
- WMD and CBRN identification and response capabilities.
- Ordnance order of battle, both friendly and enemy.
- Site exploitation/weapons technical intelligence (positive identification of ordnance and firing circuits for intelligence and tracking trends, fragmentation and crater analysis, first seen ordnance, post blast investigation).
- Protection measures/integration with base defense efforts (IED/UXO awareness training, CREW, construction of UXO pits, training of explosive ordnance clearance agent, advise on counter-IED TTPs, assist with vulnerability assessments, blast and fragmentation mitigation measures).
- Protection of personnel and property (calculating safe distances, creating protective works, RSP vs blow in place, the ability to preserve critical infrastructure that is threatened by an explosive device).
- Training capabilities of the EOD platoon and teams (UXO) awareness, counter-IED, homemade explosive awareness, IED recognition/indicators, CREW utilization, host nation EOD training, react to IED/UXO).
- Concept of Operations (CONOP) development for special programs.
- EOD support provided to SOF and other governmental agencies.
- Joint, multinational and host nation EOD capabilities.
- Support required from the BCT.
- EOD response.
- Prioritization of incidents.
- Advantages of dedicated security vs quick reaction force support.

2-25. Information gathered from the EOD platoons keeps the company command informed of the number, type, and length of missions that are being run by the teams. The platoon will also inform the company command of the types of threats that are being encountered by the teams. The company will use this information to inform the supported brigade so that they are aware of any changes in enemy TTPs in their AO.

COMMUNICATIONS

2-26. EOD teams are postured to respond to all friendly units operating throughout their AO. The EOD team maintains communications with the EOD platoon as well as the supported unit. The EOD team will communicate with their security element as well as the unit that is at the incident site, so that they can gain a better understanding of the incident site before arrival.

2-27. For every response, whether planned or emergency, the EOD team will create an EOD incident report in the EOD Information Management System (EODIMS) that may include voice, data, digital images, and video. The EOD team will send the report to the EOD platoon headquarters. A secure and robust communications system gives the EOD commander the ability to maintain awareness of current explosive ordnance, IED, and homemade explosive threats and relay the information to the supported commander.

2-28. A communications system that provides connectivity, starting from the reporting unit that calls in the EOD 9 Line report, to units operating throughout the AO and across all joint and multinational forces is vital to planning, conducting, and sustaining operations. EOD must maintain communications when its forces are widely dispersed and operating independently from their higher headquarters. EOD must be able to provide for the timely flow of information in accordance with the commander's priorities with integrated and secure communications capable of line of sight, beyond line of sight, and reach back to DOD and national level agencies to achieve unity of effort (figure 2-1).

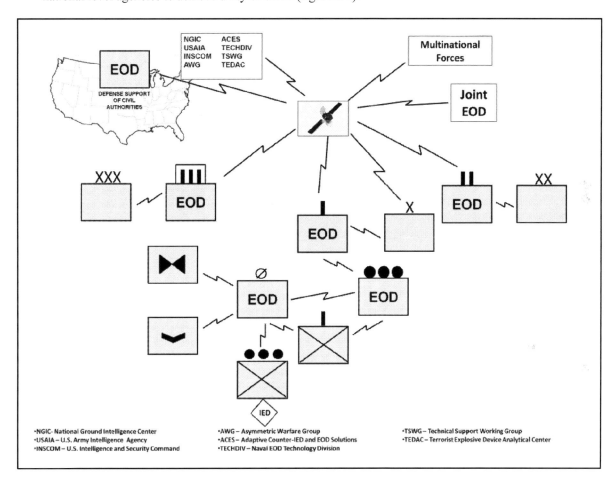

Figure 2-1. Communication architecture

2-29. The reporting of information is fundamental to EOD operations. The EOD company operations sections will be linked in with the supported units' operations section in order to maintain awareness of all missions that are in the planning or execution phases. This information will be passed down to the platoon that has responsibility of that particular AO.

2-30. When developing the EOD report it is important to use approved terminology that can be easily understood by all. The Weapons Technical Intelligence IED Lexicon is intended to encompass the broad spectrum of IED employment scenarios, the variety of IED devices, and their critical components. The lexicon was developed and approved by subject matter experts from military and civilian agencies as well as North Atlantic Treaty Organization.

2-31. The platoons maintain situational awareness of their teams, which may be geographically dispersed across the supported units AO. This is essential in to order stay informed of the missions that the teams are responding too. This information is passed up to the company so that the company can maintain situational awareness of all of the teams.

INTELLIGENCE

2-32. EOD relies on intelligence developed about an area to prepare for missions. Before a company is assigned to an AO they will gather all intelligence possible on the area, to include types of threats, enemy TTPs, number of EOD incidents, and current EOD team SOP. Some of the resources that a company can use to gather this information are to contact the EOD task force, or the EOD company they are replacing. The EOD company will utilize the supported brigades intelligence fusion cell to gather information on the AO.

2-33. The information gathered and recorded by EOD teams affect all friendly forces by identifying enemy TTPs, employment of weapons systems, types of weapons systems, and enemy targets. This information also influences friendly TTPs, identifies threat frequencies, and can affect the types of equipment that friendly forces use, such as ground penetrating radar, rollers, etc.

2-34. The EOD team is an essential part of the targeting process. Information gathered by EOD technicians by rendering safe IEDs or collecting material of interest during searches and post blast investigations directly affects the targeting process of individuals or groups involved in the IED network.

2-35. EOD forces collect, recover and exploit first seen ordnance items, WMD, CBRN, high visibility or experimental ordnance, specific IED types, unknown or homemade explosives. The information collected provides US and friendly forces with information on new or emerging enemy capabilities.

CREW

2-36. Management of the electromagnetic spectrum is essential throughout all military operations. The US military will comply with US and host nation regulations and obtain applicable authorizations before operating any spectrum dependant systems (DOD Instruction 4650.01).

2-37. The EOD company and platoon will advise on employment practices of CREW systems. EOD units will come in contact with Army, joint, and multinational forces on a regular basis. Because of the contact with other forces, EOD will have the best understanding of the capabilities of CREW systems that are in use, to include mounted and dismounted systems. When responding to an incident the EOD team is in charge of all CREW on site due to the effects on communications and robotics.

2-38. The EOD company and platoon can advise on threat frequencies used in the AO based on threats in the area. This information is gained by reports from the AO and passed on to the EOD company and platoon from the CJTF intelligence section. Information from these reports will assist commanders in determining which CREW systems work best in their AO.

2-39. Use of CREW should be a consideration when conducting operations in semi permissive environments. Before employing a CREW system, the senior EOD technician will ensure that proper authorization has been given.

PLANNING CONSIDERATIONS

2-40. Having as much information as possible before conducting either a planned operation or an emergency response is vital for mission success. Supported units need to understand the importance of including EOD in the planning process and daily intelligence update. When focusing on specific elements of an OE METT-TC should be used (ADRP 3-0).

2-41. EOD company and platoon leaders must participate in the supported brigade/battalion military decision making process and rapid decision making process to ensure EOD capabilities are integrated into the maneuver units plans.

2-42. Environment and infrastructure considerations must be taken into account. Determining the effects that EOD actions have on the surrounding area will affect the actions taken by the EOD team. These considerations will affect the priority and category of the incident.

Chapter 2

PLANNED OPERATIONS

2-43. Planned operations in support of contingency operations should include EOD due to the likely threat of explosive ordnance/IED/homemade explosives/WMD. EOD will play a large role in the planning of missions as well as the execution. EOD is one of the best sources for the supported unit when it comes to the most up to date enemy TTPs as well as the explosive threat in the area.

2-44. EOD is an integral element of mission execution that could involve, but are not limited to the following situations:
- Airfield seizure (runway clearance of UXO, IED, aircraft hazards).
- Counter WMD/CBRN.
- Capture of surface to air missile sites (man portable air defense systems, mobile and fixed sites).
- Capture or encountering of enemy ammunition supply points (exploitation and disposal of enemy ordnance).
- Global response force contingency planning.

Rehearsing

2-45. Before the execution of a planned operation, it is necessary for the company command and operations section to conduct rehearsals with the supported unit. This is necessary to ensure that the supported unit is aware of the capabilities that the EOD company can provide. The supported unit must know who can provide guidance regarding decisions that must be made regarding explosive ordnance/IED/homemade explosive and the possible effects of the team's actions.

2-46. The company must also use this time to ensure that the EOD platoons and teams have a full understanding of the mission that they are preparing for. Effective mission analysis will ensure the appropriate allocation of resources to fit mission profile.

Execution

2-47. During the execution of the planned operation, the company command and operation sections must maintain situational awareness of the actions of the EOD personnel that are taking part in the mission. If the supported unit needs guidance regarding any explosive ordnance/IED/homemade explosives related issues, EOD leadership must be present to give accurate information and have the capability to make recommendations to the supported commander.

After Action Review

2-48. After mission completion, the EOD command will participate in the supported units after action review. The EOD team will also conduct an internal after action review with the EOD command in order to disseminate lessons learned. This information will enable the command section to better prepare for future mission support. The successes or failures of the operation should be reviewed and reported to the EOD battalion.

EMERGENCY RESPONSE OPERATIONS

2-49. EOD teams are on standby to respond to any type of explosive ordnance/IED/homemade explosive incident 24 hours a day. An emergency response is typically referred to as an EOD 9 Line and is sent through the reporting unit's chain of command. Proper EOD 9 Line reporting procedures will be found in unit or theater SOP.

INCIDENT REPORTING

2-50. When the supported unit receives an EOD 9 Line they will forward it to the EOD company command post so that the company can dispatch the required amount of EOD assets to accomplish the mission. If an EOD company is not collocated with their subordinate platoons they may be informed of an EOD 9 Line response by the EOD platoon or team that is conducting the mission.

Contingency Operations

Table 2-1. EOD 9 Line

Line1	Date Time Group	131200ZAUG04
Line 2	Reporting Unit/Location	1/7th CAV 13221433
Line 3	Contact Method	F400, Sapper 6, CPT Baim
Line 4	Type of Ordnance	82 MM Mortar 1 ea
Line 5	CBRN contamination	Yes, Soldiers have blister; M8 paper confirms H
Line 6	Resources Threatened	Personnel, mine clearance equipment
Line 7	Impact on mission	Mine clearance operations are stopped
Line 8	Protective measures	Personnel evacuated to 300M; sandbagged barrier constructed
Line 9	Recommended priority	Immediate

- Line 1. Date-time group (DTG). The DTG when the item was discovered.
- Line 2. Reporting unit and explosive ordnance location. The unit designation of the reporting unit and the location of the explosive ordnance in an 8-digit grid coordinate.
- Line 3. Contact method. Provide the radio frequency and the call sign, and/or the telephone number and point of contact.
- Line 4. Type of explosive ordnance. Note the size, quantity, type of ordnance (dropped, projected, placed, possible IED, or thrown). Indicate the emplacement method and type of initiation device.
- Line 5. CBRN contamination. If CBRN is present, be as specific as possible. (For example, chemical agent monitor detected G agent at 3 bars; Soldiers are experiencing symptoms of nerve agent; excessive amount of dead wildlife).
- Line 6. Resources threatened. Report any equipment, facilities, or other assets that are threatened.
- Line 7. Impact on mission. Provide a short description of your current tactical situation and how the presence of the explosive ordnance affects your status (delayed, diverted, cancelled).
- Line 8. Protective measures taken. Describe measures taken to protect personnel and equipment (evacuated to 300M, item marked, sandbag barrier constructed).
- Line 9. Recommended priority (immediate, indirect, minor, or no threat). Ensure that the priority recommended corresponds with the tactical situation described on line 7 of the report (impact on mission). These priorities refer only to the explosive ordnance impact on the current mission. A priority of MINOR or NO THREAT does not mean that the explosive ordnance is not dangerous.

CATEGORIES

2-51. Depending on the operational tempo of an AO, an EOD team may be responding to several different EOD 9 Lines throughout the day and night. While the supported commander sets the priority of missions it is important for the supported unit to receive guidance from the EOD command and operations sections on how to best utilize EOD assets.

2-52. The EOD headquarters or operations section can advise the supported commander on the length of time or number of teams it may take to complete a mission. They can also provide the supported commander with information such as additional assets that may be required.

2-53. The supported commander should be informed of the categories of EOD missions as outlined in STANAG 2143, Explosive Ordnance Reconnaissance/Explosive Ordnance Disposal. The mission category should be established by the EOD commander, and can be upgraded or downgraded by the EOD team leader once on scene. The categories are:
- CATEGORY A. Assigned to EOD missions that constitute a grave and immediate threat. Category A incidents are to be given priority over all other incidents. Render safe and disposal operations are to be started immediately regardless of personal risk.

Chapter 2

- CATEGORY B. Assigned to EOD missions that constitute an indirect threat. Before beginning EOD operations, a safe waiting period may be observed to reduce the hazard to EOD personnel.
- CATEGORY C. Assigned to EOD missions that constitute little threat. These incidents will normally be dealt with by EOD personnel after Category A and B incidents, as the situation permits, and with minimum hazard to personnel.
- CATEGORY D. Assigned to EOD missions that constitute no threat at present.

Coordinating for Support

2-54. A variety of support may be needed in order for an EOD team to accomplish their mission. Some support can be coordinated ahead of time, while other support may be rapid requests based on the situation. The EOD command must inform the supported unit as soon as possible if they will require any additional logistical or maintainer support outside of the normal support provided.

Logistical Support

2-55. With EOD platoons and teams supporting decentralized operations the EOD company leadership coordinates with the supported units supply section so that they are able to promptly supply their platoons and teams. Platoon sergeants need to communicate with the teams so that they are aware of what the teams have on hand and be able to plan in advance for what they may need.

2-56. Types of critical items that the company will have to arrange for are:
- Supply of explosives and EOD peculiar ammunition.
- Robotics.
- EOD specific tools and equipment.
- Maintenance procedures for EOD tools and equipment.

2-57. EOD companies must integrate logistically with supported units to provide logistical support not resident within the EOD companies' organic capability, to include field maintenance and resupply.

2-58. EOD companies are assigned one wheeled vehicle mechanic. The mechanic is located with the company headquarters section and can be sent to platoons or teams if needed. EOD teams perform operator level maintenance on their equipment but may need the assistance of the supported units' maintenance section if they are not collocated with the company mechanic. The EOD company must ensure that the supported unit has the capability to provide maintenance support.

Security

2-59. Depending on the OE the EOD team may need to be augmented with additional security to establish and maintain a cordon while executing the mission. Providing EOD with a dedicated security element will expedite their response to the requesting unit. Dedicated security also allows the EOD team to build trust and a working relationship with their security element. The EOD team leader will work with the security element to determine the order of march, route to the incident, safe area location, and security formation in the safe area.

Medical

2-60. EOD Soldiers are trained as combat life savers and are capable of providing immediate life saving measures. However, due to inherent dangers of EOD operations it is preferred to have combat medics present during EOD missions. When briefing the supported unit it should be stressed that EOD teams do not have medical support.

2-61. The EOD Soldier inherently experiences greater exposure to the effects of blast overpressure and fragmentation. Due to the long term effects, leaders should continuously monitor Soldiers that are exposed to blasts and have the exposure annotated in their medical records.

EOD TEAM OPERATIONS

2-62. The EOD team is the back bone of the EOD force. On a daily basis the EOD team will respond to Army, joint, multinational, and host nation forces spread across several OE. In order to make the best decisions in support of the mission, the team must be flexible and have an exact understanding of the supported commander's priorities. EOD teams will also have contact with local nationals. The actions of the EOD team will have an impact on operations.

EOD TEAM SUPPORT TO PLANNED OPERATIONS

2-63. The EOD team brings a unique set of skills that must be properly utilized during planned operations. The team leader should brief the mission commander on the best utilization of the team. Once the EOD team is called forward to assess explosive ordnance/IED/homemade explosives the team leader will then take responsibility of all explosive related safety. The situation and type of mission will dictate what actions the team leader will take. The lead unit executing the mission needs to be aware that the EOD team leader may direct them to move out of an area in order to seek cover.

Route Clearance

2-64. Most route clearance tasks are normally not conducted under direct fire. Explosive ordnance disposal teams support route clearance by providing the technical expertise to render safe and dispose of all explosive ordnance/IED/homemade explosives that are located. Depending on the situation, EOD teams may incorporate themselves, under limited circumstances, into route clearance patrols or may respond to route clearance patrols after receiving an EOD 9 Line.

2-65. The EOD team may integrate themselves into the route clearance patrol in several ways. The EOD team leader can ride in the Mine Protected Clearance Vehicle (MPCV) in order to advise on the actions to be taken when explosive ordnance/IED/homemade explosives are encountered. The team members will ride in follow on vehicles with the EOD specific equipment and explosive charges. The team leader will keep in contact with the team members and, if necessary instruct them to perform actions on the explosive ordnance/IED/homemade explosives. In certain instances, the team leader may also find it necessary to exit the MPCV in order to perform actions on explosive ordnance/IED/homemade explosives or collect IED related material after disruption or disposal.

2-66. EOD teams may also support route clearance patrols by incorporating their response vehicle into the route clearance formation. This allows the EOD team to bring along all equipment and special charges in order to perform actions on all explosive ordnance/IED/homemade explosives. The EOD vehicle will never lead the route clearance patrol.

2-67. A good working relationship should be established between the EOD and route clearance teams. Each element must understand the capabilities and limitations of the other. The route clearance commander and the EOD team leader works together to take the appropriate action on a suspect item. There are times when it is best for the route clearance team to use the MPCV arm to investigate an area. If explosive ordnance/IED/homemade explosives are confirmed, the route clearance team should allow EOD teams to perform actions in the same manner as they would on a routine EOD response.

Air Assault

2-68. Depending on the AO, air assault missions may be necessary. During an air assault mission, the EOD team will be limited in what they will be able to bring with them in order to perform their mission. The EOD team may also be limited to the number of EOD Soldiers that are able to incorporate into the assault force. If available, the use of all terrain vehicles will allow the EOD team to expand their capability.

2-69. Helicopter landing zones may be targeted by enemy forces. During the planning process the EOD team leader should look at potential landing zones and determine if the terrain creates channeling or funneling effects. This could be an indication that there is a possibility that there will be IEDs in the landing zone or surrounding areas.

2-70. Once on the ground, the EOD team will position themselves where they are directed and assist the assaulting force until a suspect item is located. If a suspect item is located the EOD team will be called in to take the appropriate action on the explosive ordnance/IED/homemade explosives.

2-71. If there are explosive items of interest that need to be recovered, instead of disposed on site the EOD team must ensure that the items are safe to transport in a helicopter. The team leader needs to take into account the static electricity that is built up when loading and unloading from a helicopter. Coordination with the assault force commander, as well as the helicopter crew, should be part of the planning process.

Airfield Clearance

2-72. Airfield clearance is a mission that is primarily performed in offensive operations. While prepping the OE for ground forces, enemy airfields will be a major target. Destroying the enemy's ability to use their air assets and their airfields will give the US air superiority. Once ground forces have seized the airfields they will need to be repaired immediately in order to allow US and multinational forces to use them.

2-73. During airfield clearance the high usage of sub-munitions is likely and therefore is a critical planning consideration. UXO will almost certainly be present because of the amount of ordnance that may have been dropped or projected onto an airfield, and into the surrounding area. EOD teams should be positioned with the airfield seizure forces in order to immediately begin clearing the airfield as well as the surrounding area.

2-74. This can be a large operation involving several EOD platoons or companies. The use of heavy machinery may also be needed in order to move large bombs from the airfield. If possible, the UXO that is on the airfield should be moved before destroying it so that there is no more unnecessary damage that will delay airfield repair.

Cordon and Search

2-75. Cordon and search missions will often encounter explosive ordnance/IED/homemade explosives that will pose a threat to US or multinational forces. If an area has a planned operation to conduct cordon and search it is recommended that an EOD team be on standby to provide immediate assistance to the forces conducting the mission.

2-76. The EOD team that is in support can stage on the outer cordon of the area so that they can respond anywhere within the search area. Buildings within the search area may be booby-trapped. Sometimes these booby-traps will be set at the entrance to a compound, near windows, or inside the building. If the search teams locate a suspicious device or suspect a booby-trap they should immediately back away from the area and call for EOD assistance.

2-77. Once at the incident site, the EOD team may deploy the robot to perform a remote search of the area. Other methods that may be used by the EOD team in order to safe the area is the use of hook and line to remotely open gates, doors, vehicles, and any other closed compartments. During these operations, the EOD team leader must ensure that all of the search force is at a safe distance and has adequate cover in case of a high order detonation.

2-78. If explosive ordnance/IED/homemade explosives are found during a cordon and search mission the EOD team will determine if an item is safe to move to a disposal area or must be rendered safe or disposed of in place. Render safe and disposal of explosives within an urban area has the potential to cause damage to the surrounding area or may injure or kill civilians, and provide enemies with opportunities to exploit the incident for propaganda.

2-79. Disposing of explosive ordnance/IED/homemade explosives may also damage relations between the local populace and US and multinational forces. Before any render safe or disposal operation, the EOD team leader needs to seek permission from the mission commander. EOD will then ensure that they make every effort to clear the area of all civilians, US, and multinational forces before detonating the charge. The EOD team leader will also coordinate with the on scene commander to ensure that all explosive operations take place within the explosive and air clearance windows. Military information support operations (MISO) can increase the effectiveness of these efforts to obtain the cooperation of local populations, or at least their non-interference, through messages and face-to-face communication.

Dismounted Patrols

2-80. Support to dismounted patrols presents many challenges to the EOD team. EOD teams are traditionally set up to respond to explosive ordnance/IED/homemade explosives in a vehicle where they have access to all of their EOD unique tools and equipment. This is not possible during a dismounted patrol.

2-81. Selection of tools and equipment for a dismounted patrol is the most important part of the planning process. Obtaining intelligence and information on the patrol area will aid in the selection of tools and equipment for the EOD team. Every EOD Soldier will need to carry a combat load, similar to the other Soldiers on the patrol. In addition to the combat load the EOD team will also carry their EOD tools and equipment, as well as explosives. The EOD team leader will be the deciding authority on what tools and equipment will be taken on the mission.

2-82. Backpack electronic countermeasures (ECM) should be incorporated into the patrol. The use of ECM during these patrols will require Soldiers to carry a significant additional weight. The EOD team may be required to spread load required EOD tools throughout the patrol. EOD commanders will advise the maneuver commander of the value of ECM protection.

2-83. For a short patrol the EOD team may be able to carry a lightweight robot that will give them the capability to remotely place an explosive charge, or disruptor next to a suspect device. The robot will also allow the team to remotely search an area where they suspect a device may be placed. Lightweight robots do not have the same capabilities of a MK 1 or MK 2 Man Transportable Robotic System (MTRS. Supported units need to understand the capabilities and limitations of these smaller robots so that they do not expect the robot to be able to perform the same functions as the larger EOD robotic platforms.

2-84. If a robot is not available, the use of hook and line provides the EOD technician the best chance to remotely place a charge on or near a suspected device. Hook and line procedures can be difficult due to terrain. An entire hook and line kit is too large to transport on a dismounted patrol. The EOD team should ensure that they select the proper pieces of the kit they are comfortable using. This may be different for each team. Before and during a deployment the EOD team must become proficient in using hook and line to place disruption charges on or near explosive ordnance/IED/homemade explosives.

2-85. Every EOD Soldier on the patrol should have drop charges already made up and ready to use in the event that they have to take immediate action on a device. In the event that an EOD Soldier, or a Soldier that is close by finds an IED that is set to fire remotely or by command, the drop charge can be placed and the area surrounding the IED evacuated. During the planning phase of a dismounted patrol the use of drop charges should be briefed so that everyone understands when the use of the drop charge is appropriate and the proper actions to take when one is used.

2-86. Before departing on the dismounted patrol, the EOD team leader needs to inform the Soldiers what actions they should take if they find suspect explosive ordnance/IED/homemade explosives. Battle drills and rehearsals will ensure that all personnel on the patrol will know what to do when a threat is encountered. These actions will depend on the terrain that the patrol is moving through and the goal of the mission. The team leader should also ensure that the EOD team is spread throughout the patrol so that each EOD Soldier is at a safe distance from the other EOD Soldiers in the case of a detonation.

2-87. Lightweight metal detectors have proven to be invaluable tools during dismounted operations. Several Soldiers using metal detectors should be spread out to cover as much area as possible during the patrol. If a Soldier carrying the metal detector locates something suspicious they will alert EOD and warn the other members of the patrol. EOD will then take the appropriate actions to confirm the presence of an IED or determine if it is safe. Metal detectors can also be used to locate metallic signatures in walls and doors. The presence of metal could be a sign that there is an explosive device hidden in the area.

Chapter 2

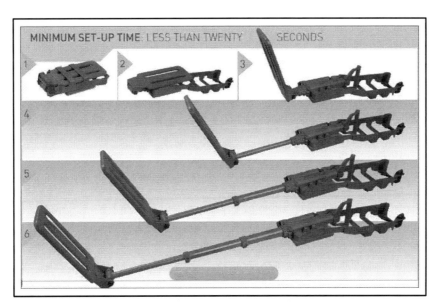

Figure 2-2. Lightweight metal detector

2-88. If the patrol finds a weapons cache or IED that is too large for the EOD team to dispose of, the area should be marked and additional assistance requested. The EOD team on site must make sure that they clear the area and make an accurate determination of what assets they will need in order to dispose of the cache or IED. Depending on the mission, the patrol will be told to standby and wait for assistance, or mark the area and continue with their mission.

Weapons Buyback/Small Rewards Programs

2-89. Removing weapons and ammunition from the local populace is essential to ensure the safety of the local populace. Additionally, these programs are used to equip and supply host nation security and police forces for training and operations. These programs have the potential to be very dangerous to those that are tasked with collecting these weapons. EOD is often tasked with ensuring that the weapons and ammunition that are turned in is safe. These missions may be conducted in support of civil military operations and other governmental agencies. MISO messages can provide specific instructions to populations about how to participate in the buyback program to better ensure the safety of all involved.

2-90. Before supporting a weapon buyback program, EOD should inform the supported unit of what capabilities that they will be able to provide. The EOD team leader will make recommendations to the weapons buyback program commander on protection issues. Before allowing a weapon to be brought into the weapon buyback area it should be thoroughly checked to ensure that it is in a safe condition. This check can be performed by the EOD team by inspecting the weapon visually or using an X-ray in cases where they are unable to visually confirm its state.

2-91. If a weapon is deemed to be in an unsafe condition, the EOD team leader will clear the area and transport the weapon to a pre-designated safe disposal area. The safe disposal area should be identified in the planning stage of the mission. It should be in an area that is able to withstand a high order detonation. People in the area need to have a safe place to take cover in case it is necessary to dispose of a weapon on site.

2-92. Weapons that are brought to be sold back must be checked to ensure that the weapon is real and in a working condition. Some weapon systems can bring a large amount of money. Because of this, someone may attempt to make a weapon seem as if it is in working condition.

Training and Partnership

2-93. The training and partnership of foreign EOD forces is an ongoing mission that Army EOD continues to participate in. EOD leaders should ensure proper authorization has been given by the approval authority prior to training foreign EOD personnel. When given the mission of training and partnership the mindset must begin before the deployment. Every Soldier that will be involved with the mission must be properly trained in the culture, language, customs and courtesies of the forces that they will be partnering. This training will pay off once deployed and will help avoid misunderstandings.

2-94. This mission requires extensive logistical support which may include additional personnel, vehicles, security, equipment, and linguists. The EOD unit assigned to the mission should take into account these extra requirements that may be needed.

2-95. Training and partnership is a very dangerous mission. When operating in these conditions both the Army and the local nationals should work together as one unit. The EOD Soldier must build relationships with those that they are training so that trust is established. Even with trust established every Soldier must stay alert and aware that those they are training and partnering with may try to harm them.

2-96. Prior to deployment, the EOD unit should ensure that they have personnel trained by the Guardian Angel program. While conducting training and partnering, these "guardian angels" will provide security for the EOD Soldiers.

2-97. Before training and partnering with local nationals, the methods that will be used must be discussed with those that are going to be trained. Host nation forces will have varying levels of certification and must be properly vetted before training and partnering takes place. The EOD Soldier must understand that some countries do not respond well to the type of training that may be conducted in the US. Trying to get a point across in English while yelling may not translate into the local dialect. This does not mean that you cannot be firm when working with the trainees. The trainees must understand that the Soldier is in charge and that no actions should be taken until receiving instruction.

2-98. When training foreign nationals in the use of explosives the trainer must pay close attention to what the trainee is doing. In many cases the Soldier will be relaying instructions and corrections through an interpreter. That means that whatever the Soldier is saying could come out a different way once it is interpreted. It also means that there will be a delay in the message reaching the trainee. It is important for the EOD Soldier to be able to recognize when a potential mistake will be made so that it can be corrected before it happens.

2-99. Some corrections must be made immediately and the EOD Soldier may have to use a hands on approach to stop an unsafe action. This should be done in a way that does not disrespect the trainee. The EOD Soldier should also learn certain words in the local language such as stop, no, good job, move, and any other essential words that will aid in training. The Soldier should ask the interpreters how to say these words and any other phrases that may be important.

2-100. While training and partnering in combat, the US EOD team will have to accompany the host-nation EOD team on responses. The US EOD team should pay close attention to how the response is being conducted and be ready to step in if the host-nation team is making life threatening mistakes. The US EOD team should make every effort to observe and stay out of the view of those that the host-nation EOD team is responding to. This will build self confidence in the capabilities of the host-nation EOD teams.

EOD TEAM EMERGENCY RESPONSE

2-101. EOD is ready to respond at all times. Once an EOD 9 line is received the team will immediately depart their area to meet up with the security team. If the supported unit has provided EOD with a dedicated security element, the response time will be greatly reduced. Being able to quickly respond to requesting units will enable that unit to continue their mission. It also prevents the finding unit from spending too much time in a vulnerable position.

2-102. If the EOD team does not have dedicated security, response time will be delayed. The requesting unit may be tasked to provide security for EOD or the supported unit will put together a security element to escort EOD to the incident site.

2-103. When an EOD team departs the base to respond to an EOD 9 line they should be prepared to respond to multiple calls without returning to base. Ensuring that the team has the correct amount and type of EOD tools, equipment, and explosives will allow the team to respond to all types of missions that they may be called to.

EOD 9 Line Report Received

2-104. When a unit locates suspected explosive ordnance/IED/homemade explosives they are to immediately evacuate the area and form a cordon. This cordon should be at a distance that provides protection from blast and fragmentation from the explosive ordnance/IED/homemade explosives in the case of a high order detonation. If possible, mark the suspected explosive ordnance/IED/homemade explosives and provide a good description of the item for the EOD team.

> **WARNING**
>
> **Finding units should not place themselves in more danger in order to mark or get a good description of the suspected threats.**

2-105. When an EOD 9 Line is received it will be passed down to the highest level of EOD command in the AO. This could be the EOD battalion, company, platoon, or in some cases it will go straight to the team. The Soldier that receives the report will ensure that it has all of the appropriate information before providing it to the EOD team on duty.

2-106. If the EOD company receives the EOD 9 line, they will inform the EOD battalion of the details of the report and dispatch the EOD team. If the platoon receives the EOD 9 line, they should inform their parent company and dispatch the EOD team. If the team receives the EOD 9 line they will immediately inform the platoon as they are departing the base.

2-107. Multiple EOD 9 lines may be received at the same time. Depending on the supported commanders priorities the EOD team will be informed of what order they should respond to the calls. EOD Teams should advise the supported commander regarding the best way to respond to support his priorities and complete the missions. If there are multiple teams able to respond then there may be more than one team being sent on response. In the case of multiple EOD teams responding at the same time there will need to be multiple security elements.

2-108. While the team leader is ensuring that all appropriate information has been received, the team members should be rechecking the response vehicle to ensure that they have all tools, equipment, and explosive charges to accomplish the mission and any other follow on missions that may be received.

2-109. The team leader should check the grid coordinates on the EOD 9 line to see if there have been explosive ordnance/IED/homemade explosives incidents in the same area. If there have been, what kind were they? If IEDs have been placed in the same general area, the team leader should know what types of setups have been used in the past. The team leader should also know of previous safe areas that have been used and what routes have been taken when responding to the same area. The team should try to vary their routes and safe areas as much as they can so that the enemy has a hard time predicting where to place secondary devices.

Movement

2-110. The EOD team will meet up with their security element at a predetermined staging area. At the staging area the EOD team leader, truck commanders, and security element leader will discuss their plan for movement to the site. This should include the route to be taken, order of march, vehicle spacing, ECM use, and actions on contact. The team members will ensure that they have established communications and will trouble shoot any problems that may come up before leaving the base.

2-111. Road conditions can make travel to incident sites very difficult. If responding to a unit that is dismounted or a tracked unit, the EOD and security vehicles may not be able to make it all of the way to the incident site. In this case the requesting unit needs to provide a link up point in an area that EOD and their security can reach. Once at the link up point the EOD team and some members of their security may need to dismount or load into one of the tracked vehicles in order to make it to the incident site. In these cases most of the security element should remain with their vehicles in order to provide security.

2-112. Other road conditions of concern are bridge crossings, driving on canal roads, and unpaved surfaces. All of these situations make it very easy for an IED to be concealed. IEDs and firing devices can be easily concealed in unpaved surfaces, under bridges, or dug into the side of a canal road.

2-113. As your convoy nears the outer cordon that has been set by the finding unit, either the security element leader or the EOD team leader will establish communications. While in route, the finding unit may have had to change their safe area due to a threat in their original area. Once communications have been established ask if there have been any changes at the incident site since they sent in the EOD 9 Line.

2-114. As the convoy comes into view ensure that you identify yourselves as EOD and approach the area slowly, constantly scanning for IEDs. If it is dark then the EOD convoy may want to turn out their lights so as to not illuminate the finding unit.

2-115. Once you arrive at the safe area of the finding unit the first thing that should be done by both EOD and their security element is to search the area for IEDs. This includes conducting 0, 5s and 25s (Figure 2-3). EOD will ensure that their security knows how to conduct a thorough safe area search, to include checking under vehicles. As soon as this is done the EOD team leader will locate the on scene commander in order to verify the information that has been received and get an accurate description of the location of the possible device. While the team leader is questioning the on-scene commander the security element and the EOD team should keep their trucks running.

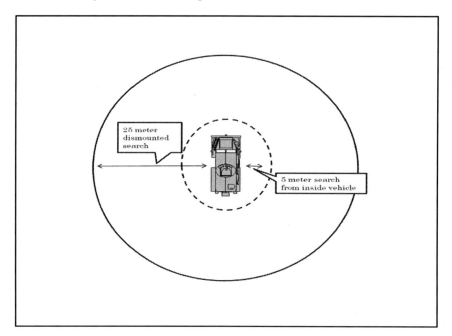

Figure 2-3. 0/5/25 meter check

2-116. At this time, the team leader needs to ensure the on scene commander understands that they will maintain their cordon until EOD declares they are mission complete. EOD security is for internal protection of the EOD team.

2-117. Based on the information received, the EOD team leader should have a tentative plan of attack established. If the robot is going to be used the team leader may instruct the team member to ready the robot while the team leader collects additional information so that it can be quickly deployed when needed.

2-118. If the finding unit is using a different type of ECM than the EOD team, they will need to deconflict any potential problems the two systems have. The EOD team should know if there are any types of ECM that may affect their communications equipment or robotics. The EOD team leader is in charge of all ECM at an EOD incident site.

2-119. The team leader will make the decision to stay where they are or move in closer to the suspected device. If EOD moves in closer they will make it clear to the finding unit that they are to maintain their outer cordon and not leave until the team leader has turned the scene back over.

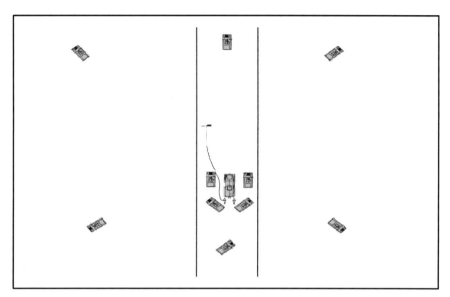

Figure 2-4. Inner and outer cordon

Note. Inform the finding unit to watch for any suspicious people in the area. Anyone in the area that has a phone or video recorder could be watching the device. The age and sex of the person does not matter. If the finding unit locates anyone that seems suspicious they should immediately inform the EOD security element.

Questioning Techniques

2-120. The information that is received from the EOD 9 Line may not be very clear or accurate. While you may begin to formulate your plan based off of the initial information, be prepared to change that plan once on scene. If possible, ask to speak to the person that spotted the suspicious item. It is best to get the information directly from the source. The source could be a local national or a civilian that may not speak English. Use your judgment in those cases as to who the information should come from.

2-121. Before the EOD team leader begins the questioning, allow the source of the information to tell exactly what they saw. By not asking direct questions at this point the source of the information may provide valuable information that the EOD team may not ask for. Once the source has finished telling their story the EOD team leader may ask direct questions to ensure the EOD team, security element, and finding unit has all possible information.

2-122. Ask direct questions such as:
- What did you see?
- Where is it located?
- Where were you when you saw it?
- Did you see a triggering device such as a pressure plate, antenna, or command wire?
- Did you see explosive ordnance/IED/homemade explosives or a container that could contain a main charge?
- Were you able to mark it in any way?
- Were you able to get a picture of it?
- Can you draw a picture of what you saw?

2-123. Depending on the situation, you may want to take the individual that spotted the suspected device in the EOD vehicle and move into the inner cordon with the team. That will enable the individual to help guide the robot operator to locate the device. This is especially important at night, or if the device is well concealed into the environment.

2-124. If the EOD team is responding to a post blast incident the team leader should question everyone that witnessed the explosion. If the IED hit a vehicle and the occupants are still on scene and able to speak, ask them what was the last thing they saw before the explosion. Something out of place may catch a Soldiers eye seconds before the explosion. Also ask if anyone on scene has any video of the incident.

On Scene Security

2-125. The EOD security element is there to provide security to the EOD team. While on an EOD response, the team leader and team members will be concentrating on their job. This may leave them vulnerable to small arms or sniper fire. Throughout a response, the security team should be scanning the surrounding area for any signs of a potential attack or potential IED triggermen or spotters. The security team leader should keep the EOD team leader informed of any suspicious activity in the area.

2-126. Depending on the layout of the area the security element will form their trucks around the EOD truck in order to provide the best protection. In an area where they are likely to take small arms and sniper fire a tight wedge formation at the back of the EOD truck will provide cover for the EOD Soldiers when they are on the ground.

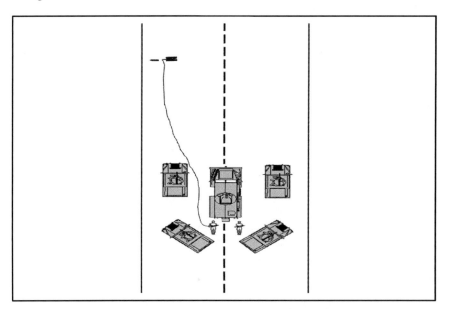

Figure 2-5. Inner cordon with wedge formation

2-127. In an urban area the security element needs to be very aware of possible enemy firing positions. It can be very difficult to provide security in an urban area and it may require that some of the security element dismount. It may also be necessary for a security Soldier to remain close to each of the EOD Soldiers in order to keep a close watch around them.

2-128. When establishing a safe area in urban terrain do not park where you are exposed to alleys or roads. If you are in an area where there are alleys or roads it may be necessary to have a security truck pull up to cover the area.

Chapter 2

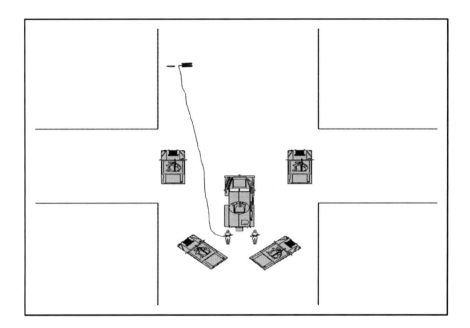

Figure 2-6. Inner cordon in urban area

2-129. The team leader should be aware of the environment that the safe area is being established in. Some areas may not be able to withstand a high order detonation in the case that the device explodes. Parking close to buildings that have been damaged or have windows that could break may cause harm to the Soldiers in the cordon. The team leader should also try to recognize if there are any gas or power lines in the area that may be damaged. Every effort will be made to protect civilian infrastructure.

2-130. The device may be located in an area that will cause the EOD team and their security to get very close. In these cases be sure to warn the security team to have their gunners keep their heads down throughout the call. When outside of the truck in these situations the EOD technicians should have minimum exposure to the device.

EOD Team Responsibilities On Scene

2-131. The EOD team leader is the operational authority for all explosive related issues within the inner cordon during an EOD mission. This is to ensure the safety of the EOD team and all others in the surrounding area. No personnel or equipment, not specifically authorized by the EOD team leader will enter or operate within the inner cordon. The EOD team leader will insure the on-scene commander is aware of all security requirements, the role of the EOD team leader at the incident site, and is notified prior to any controlled detonations during EOD operations.

2-132. The EOD team leader will advise the senior commander on the ground on the use and positioning of all ECM and communications equipment on site during the conduct of all EOD missions. ECM and communications equipment which are not utilized correctly will result in the reduction or elimination of coverage provided by ECM devices within the inner cordon or within the vicinity of the EOD response.

2-133. The EOD team is responsible to confirm and neutralize, if necessary, the explosive device that they were called out to investigate. While on site, additional suspected explosive devices may be located. The EOD team is also responsible to confirm and clear those as well.

2-134. Once a device has been rendered safe or disposed of, the EOD team is responsible for visually confirming that the hazard has been eliminated. If anything is left to recover, the team will gather device components and forensics samples for follow-on exploitation at an exploitation facility.

2-135. When the explosive device is rendered safe and explosives are still left, the EOD team will dispose of the explosives on site if the area can withstand a high order detonation. If not, the EOD team will transport the explosives to a safe disposal area or back to a safe holding area until they are able to destroy it.

EOD Team Actions On Scene

2-136. The type of call that the EOD team has been called out for will determine the actions taken. In all cases the EOD team needs to be cautious and continually search for IEDs in the area. By placing an IED, UXO, or any suspicious item that will cause a unit to stop in a predetermined area the enemy will have the opportunity to attack. The EOD team needs to complete the call in a quick but safe and thorough manner.

Improvised Explosive Device

2-137. When the EOD team has cleared and established a safe area, and the security element has been set, the team will begin their actions on scene. If responding to a suspected IED, the team leader may elect to send the robot downrange with an explosive charge or disruption tool to the area where the suspected IED is reported to be. The robot driver may find it easier to drive the robot backwards so that the shock tube is not tangled in the tracks.

2-138. Once the device has been located, the team leader can tell the robot operator to either set the charge to blow the IED in place or to set the charge in a manner that will separate the IED components without causing a high order detonation. If the robot operator cannot locate the IED, they may set the charge down so that they are able to drive the robot around the area to locate the device. Once the device has been located the robot operator may then pick the charge back up and place it. If using these techniques at night it may be necessary to attach an infrared chemical light on the charge to make it easy to find once it has been dropped. A chemical light may be attached to the robot to help the robot operator maintain situation awareness.

2-139. If the IED main charge has not been located but a device for making, breaking, or changing a connection in order to trigger the device, such as a command wire, pressure strip, battery pack, or telephone base station has been, the team leader may have the robot operator use the robot to take the IED apart with the grippers. These operations can be very difficult but are helpful in recovering components and forensic material for exploitation purposes. These techniques are also useful because they will help in finding the main charge of the IED. If using the robot to take an IED apart it is possible that the robot operator will trigger the device. Be sure to inform both your security element, and the finding unit that they need to maintain cover throughout the call in the case of a high order detonation.

2-140. When using the robot to place a charge or pull components from an IED, it is best to have the arm extended. This will provide the most leverage for pulling heavy objects. Having the arm extended will also keep the robot body as far from the main charge as possible. This could possibly limit the damage to the robot in the event that the IED detonates.

2-141. After the IED has been rendered safe or disposed of by detonation, it may be possible to recover components and forensic material using the robot. Before recovering any components, the team leader must determine if the benefit outweighs the risk. If the robot operator is able to recover any components and forensic material and bring it back to the safe area the team leader should check it to make sure that it is explosively safe. IED switches, such as telephone base stations, have been booby trapped and could cause harm if not properly screened and evaluated for hazards. Before placing the recovered IED components into the truck the team leader needs to remove any batteries and ensure that there is not a blasting cap still attached.

2-142. The EOD team should recover all IED components, to include containers. Command wire that is used should not be left in place. Tracing the command wire out to the firing point can provide the EOD team with valuable information on enemy TTPs that they can relay to the supported unit. If the team is tracing the command wire out they should never walk directly along the path. When recovering the command wire it should be done remotely in case it is booby trapped.

2-143. When recovering IED components for forensic purposes, be sure to wear latex gloves. Bag and mark the items appropriately so that the evidence collected is not compromised. If the IED main charge is homemade explosives, a sample should be taken so that the chemical makeup of the explosives can be tracked. Sample size may be dictated by theater policy. (ATTP 3-90.15, *Site Exploitation Operations*)

Unexploded Explosive Ordnance

2-144. Unexploded Explosive Ordnance response calls should be treated in the same way that an IED call is treated. The EOD team that is responding should assume that the UXO is either booby trapped or set up as an IED until they can rule out that threat. When arriving at an UXO site, treat the arrival the same as you would on any other response call.

2-145. Determine the condition of the UXO and whether the area is able to withstand a high order detonation. If the supported unit allows you to destroy the UXO in place then do so unless the UXO is determined to be of intelligence value. If the UXO is in an unsafe condition and cannot be moved, the team leader will determine whether to perform RSP or dispose of the UXO in place. Before destroying the UXO in place be sure that everyone in the blast and fragmentation zone has adequate frontal and overhead protection.

2-146. All bases should have a dud pit where explosive ordnance that has been recovered from weapons caches, or turned in should be stored until EOD can destroy them. The dud pit should be located in a place that is far enough from living and work areas that there will be minimal damage in the event of an explosion. In cases where a base is not large enough for a dud pit, a place outside the gate, but within site of the gate guards, should be established. EOD teams should advise on where to place dud pits and should check them regularly.

2-147. There are times when items are placed in dud pits that are in a dangerous condition. EOD Soldiers should continually inform US, multinational, and host nation forces on what items should never be brought into a dud pit, such as any explosive ordnance that is in an armed condition or damaged. If someone tries to turn in UXO that appears to be in an unsafe condition they should be stopped and told to place the item away from the entrance to the base. At that time the operational area security can call EOD to dispose of the UXO appropriately.

Weapons or Explosive Precursor Cache

2-148. In the event that a unit locates a weapons cache they should immediately back out of the area, set a cordon and call EOD. A weapons cache may have booby traps, landmines or IEDs emplaced around it (if landmines are confirmed, engineers may be called to assist). When EOD arrives on scene they should receive a detailed brief of what was seen, where it is located, and the path in and out that the finding unit took.

2-149. If possible the robot should be used to approach the cache. If nothing suspicious is located by the robot operator then one EOD technician should approach using a metal detector and searching for landmines and IEDs. The outside of the cache should be cleared before entering the area to remove the contents.

2-150. Once the outside of the cache is deemed to be safe, one EOD technician can enter the cache, constantly checking the area for booby traps. The EOD technician should determine how best to remove the contents of the cache. Depending on the contents, the items may have to be remotely moved from the area. This can be a time consuming operation and the EOD team leader should keep the finding unit informed of their progress.

2-151. A weapons cache may also be buried in the ground. This can make recovery of the items very dangerous because the EOD team may have to dig. The team leader can request support from the finding unit if it is deemed safe.

2-152. If nothing in the weapons cache is worth collecting for exploitation purposes then the EOD team may destroy the weapons on site. If the area cannot withstand a high order detonation then the items will have to be transported to a safe disposal area. Some weapons caches can be very large; in this case the supported unit must supply EOD with a vehicle capable of transporting the items to a disposal area.

Post Blast Analysis

2-153. Information collected by EOD teams conducting post blast analysis shows enemy TTPs that are working. The placement of the device and the location of the hit on the vehicle will show whether the enemy has figured out how to counter friendly TTPs. The type of firing device will tell us if ECM is effective. Damage to vehicles or personnel that have been hit will tell us what type of main charge that the enemy is using.

> *Note.* Reports generated by EOD teams have been seen at the highest levels of the government. The EOD team leader and EOD leadership will ensure that these reports are informative and accurate before submitting them higher.

2-154. Post blast scenes can be very chaotic, especially when civilians have been targeted by large IEDs, such as a vehicle-borne IED. While in route to the incident site, EOD should inform the unit that is setting the cordon to ensure that all personnel are evacuated from the scene. The unit securing the area should also prevent any tampering with the scene because this could destroy evidence.

2-155. The EOD team should treat a post blast analysis the same way that they would any other emergency response. The team needs to be cautious of other IEDs in the area. First responders, such as EOD, are often targeted when responding to a post blast incident.

2-156. Most post blasts conducted during contingency operations will be tactical post blasts. Due to the limited time that an EOD team will have to conduct the investigation, the team leader must ensure that everyone in the security detail and the EOD team members understand their role. When planning for the post blast the team leader must consider cordon distance, security emplacement, safe area and route selection.

2-157. The team leader will ensure that everyone knows what actions to take if they begin receiving sniper, direct or indirect fire. Also instruct everyone on actions to be taken in the case that a secondary IED is located or there is an IED strike in or near the cordon.

2-158. Non explosive hazards may be present within a blast scene. The EOD team must consider the following hazards:
- Burning buildings or cars
- Building structural integrity
- Biological hazards
- Blood and tissue
- Utilities (water mains, sewer mains, natural gas/propane)

2-159. Explosive hazards that must be considered are:
- Secondary or tertiary devices
- Homemade explosive precursors (inhalation, contact burns, low oxygen levels)

2-160. The team leader will determine what type of approach to use when arriving on scene. If EOD is near the area when the blast occurs there may still be friendly personnel that need to be evacuated. The team leader will determine on scene whether they will take the risk to approach the area on foot in order to clear a path to evacuate injured personnel.

2-161. Another option that the EOD team has when conducting a post blast analysis is to use the mine protected vehicle in order to conduct a hasty analysis of the crater. The EOD team may elect to remain in the vehicle and drive by the crater. This technique can be used if it is determined that the post blast is a low priority mission.

2-162. If the team arrives after the injured personnel have been evacuated the team should deploy the robot to check the area for secondary or tertiary devices. Any IED components that are found should be remotely moved in case it is booby trapped. Evidence will be collected and properly bagged for further exploitation. A soil sample from the blast seat, the size of the crater, or fragmentation found will aid the EOD team in determining the type of explosives as well as the explosive weight of the IED.

2-163. The team leader will determine the best method to search the area. There are four different patterns that will ensure the scene is thoroughly searched. The circle pattern can both start at the perimeter of the search area and then move into the center or start at the center and move to the outside of the search area.

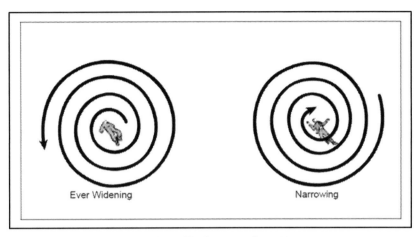

Figure 2-7. Circle search pattern

2-164. If the search area is large, the team leader may elect to conduct a search by using the strip/lane pattern. This consists of the EOD team getting on line and searching lanes that are roughly one to two meters apart. This method is beneficial when there is limited time on scene.

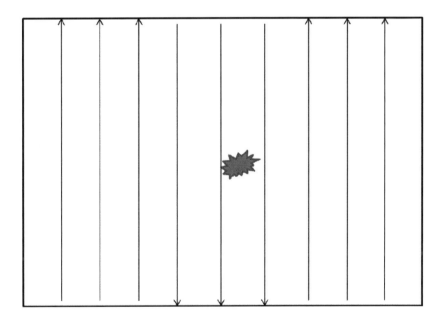

Figure 2-8. Strip/lane search pattern

2-165. If time at the scene permits, a method to conduct a very thorough search of an area is the zone search. To conduct a zone search the team will sketch out the scene and assign and number zones. As the specific zone is searched the evidence that is collected will be tagged with the zone number that it came from. This method is very effective for searching houses or buildings with several rooms.

Contingency Operations

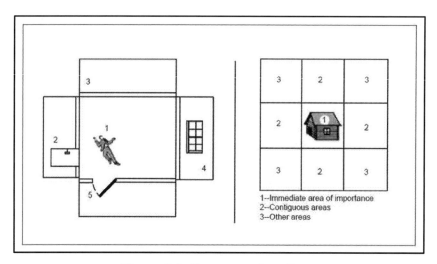

Figure 2-9. Zone search pattern

2-166. Another method that is effective for both indoor and outdoor searches is the grid pattern. The area will be divided and the search team will start at the corner and begin following the pattern to ensure that each grid is meticulously searched. This method ensures that each of the grids has been looked at twice.

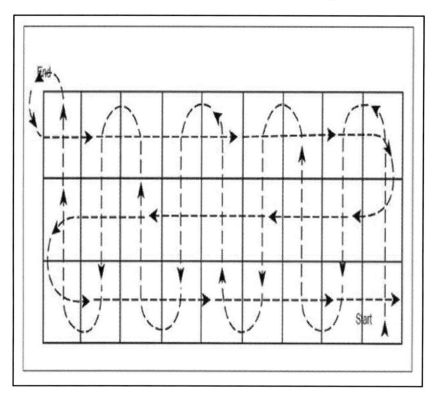

Figure 2-10. Grid search pattern

2-167. Thorough documentation of the scene is crucial in conducting a post blast analysis. Pictures, video, or sketches of the scene and surrounding area will allow readers of the post blast report to have a better understanding of the environment that the enemy is using. Notes taken during the post blast will benefit the team leader when filling out the report.

Chapter 2

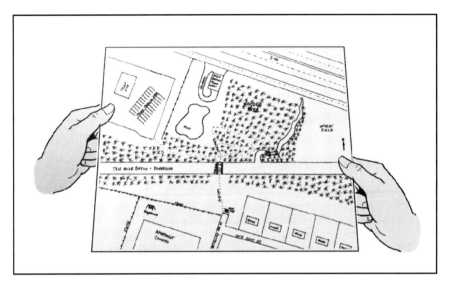

Figure 2-11. Scene sketch

2-168. Scene sketches should make an effort to highlight areas around the blast site. These areas could have been used by the bomber as either aiming points or hiding places that could be used to trigger the IED or video the attack. The EOD team should make note of the following when sketching the scene:

- Buildings in and around the site.
- Landmarks.
- Location of significant events.
- Reference points.

2-169. The EOD team must be careful when handling evidence. The careful handling of evidence is not only to ensure that the team is safe, it is also necessary to limit cross contamination of the site or evidence that is collected. All team members must wear latex gloves during the search. Be sure to check the gloves throughout the search in the case that they are cut by sharp objects. The wearing of gloves does not necessarily mean that fingerprints will not be destroyed. All team members should handle collected evidence in a way that limits the destruction of any potential fingerprints.

2-170. EOD teams will have to handle weapons as part of a tactical post blast. When handling weapons the team must take the following into consideration:

- Place weapon on safe prior to handling.
- Handle areas least likely to hold fingerprints.
- Do not handle smooth surfaces of the weapon.
- Limit the handling of magazines.
- Do not remove rounds from the magazines.

2-171. The EOD team will also be exposed to Deoxyribonucleic acid (DNA). Human or animal DNA can be a hazard to the EOD team collecting the evidence or working at a post blast scene. If not properly collected the DNA could be a hazard to lab personnel that are exploiting the evidence. The following should be considered when working with DNA:

- Potentially infectious.
- Significantly degraded.
- Biological evidence to be air dried as soon as possible.
- Do not write directly on any evidence.
- Control of access to the scene.
- Fuel powered tools present potential contamination.
- Wear a fresh pair of latex gloves.

- Use clean tools (if clean tools are not available it should be noted in the report).
- Place in clean unused containers.
- Package liquid samples to prevent leakage.

2-172. At some point in the investigation the EOD team leader should question any available witnesses, to include local nationals. When questioning the witnesses be sure to take into account that these people have either been involved with or been witness to a traumatic event which may have altered their perception.

2-173. After evidence has been collected the team will begin to close down the site. Before leaving the area the team leader must back brief the on scene commander and inform them that they may collapse their cordon.

Scene Turnover

2-174. After the EOD team has completed their mission they will turn the scene back over to the requesting unit. The team leader should brief the on-scene commander on what they encountered. If the team has rendered safe an IED and recovered components it may be beneficial to the finding unit to see what was recovered and get a brief description of how the IED was set up.

> *Note.* Never guarantee that an area is explosively safe. Let the requesting unit know that you have rendered safe, or disposed of all hazards that were found. Tell them to proceed with caution.

2-175. EOD teams will respond to the same units throughout their deployment. The scene turnover is a good time to build a rapport with the units that are operating in the area. This relationship may prove to be useful in future responses because requesting units will know what EOD needs them to do when they request EOD support. After back briefing the on-scene commander the requesting unit may collapse their cordon and carry on with their mission.

Follow On Calls

2-176. When an EOD team leaves to respond to one call they should be prepared to stay out for follow on calls. A follow on call is any EOD 9 Line that comes in before a team is able to return back to base. When the supported unit receives an EOD 9 Line, they will forward it to the EOD security element. The security element will copy all information down and give it to the EOD team leader.

2-177. A follow on call may be received that has a higher priority than the current call that EOD is responding to. If that is the case, the EOD team may have to stop working and immediately respond to the higher priority call. In this case, the requesting unit of the call that EOD has responded to may have to maintain security at the site until EOD is able to return, or they may mark and leave the site if ordered to.

2-178. EOD may receive several calls throughout the day and night. Ensure that there is a clear understanding what priority these calls have so that they are done in the correct order. The team leader, and security element leader, should also take note of where the calls are located. If operating in a large area of operations, the calls may be spread far apart. The EOD team may recommend a change in priorities based on location to the supported unit.

2-179. If an EOD team and security element has responded to several calls without returning to base, they may be running low on supplies. If this happens and they are close to the base they may have to return to base to resupply before responding to the follow on calls that are still waiting. If a team has been out for too long, or if there are several calls that have a higher priority, the supported unit may request a second or third EOD team be dispatched if there are teams available.

Chapter 3
Defense Support of Civil Authorities

Defense support of civil authorities operations encompass all support provided by the components of the Army to civil authorities within the US, its possessions and territories. This includes support provided by the regular Army, Army reserve, and Army National Guard (when the Secretary of Defense, in coordination with the Governors of the States, elects and requests to use those forces in title 32, United States Code, status). Army forces frequently conduct DSCA operations in response to requests from federal, state, local, and tribal authorities for domestic incidents, emergencies, disasters, designated law enforcement support, and other domestic activities. DSCA operations are related to stability operations and foreign humanitarian assistance, but while these missions are conducted overseas, DSCA operations are conducted within the US, its territories and possessions, in support of the US population. As IED/WMD threats increase within the homeland, EOD forces will offer valuable expertise in response to these threats.

GROUP AND BATTALION

3-1. While most support to civil authorities request for assistance will be conducted by the EOD companies, the EOD group and battalion play a vital role in ensuring that support provided by EOD is prompt and suitable for the assigned mission.

LEADERSHIP

3-2. Commanders gauge unit readiness for DSCA missions by assessing proficiency in three warfighting functions: mission command, sustainment, and protection. The requirement to deploy into a domestic OE, often with little warning, and to operate with joint and interagency partners requires mission command that can adapt systems and procedures for a noncombat, civilian-led structure (ADP 3-28). Group and battalion commanders must ensure that all EOD units are equipped, trained and ready to respond to requests from civil authorities. The group and battalion must also ensure that all EOD Soldiers understand what capabilities may legally be used when providing EOD support.

3-3. EOD forces are in a constant rotation either preparing to deploy, deployed, or redeployed. DSCA is a mission that is done in conjunction with the normal mission command battle rhythm. EOD leaders recognize the strain that this puts on Soldiers who are training to execute their overseas mission. The group and battalion leadership must ensure that DSCA is executed in a manner that allows EOD forces to properly prepare for deployments.

3-4. EOD corps/division/brigade/battalion staff officers and NCOs will provide support to DOD and government installations with vulnerability assessments to determine the threat to the installation if attacked with explosives. The EOD company commander should make this information available through a capabilities brief, which will include communications, intelligence, and CREW used by EOD forces in the homeland.

COMMUNICATIONS

3-5. The EOD battalion is responsible for ensuring that EOD companies have the training and resources required to maintain communications with supported military organizations and units. The EOD battalion is responsible for providing training and operational radio frequencies and call signs monthly or as required to EOD company communications representatives. The EOD battalion should work closely with the

Chapter 3

installation provost marshal office (PMO), range control, and spectrum management sections to ensure that EOD companies are fielded compatible radio systems to communicate with all installation supported agencies in support of emergency response operations. When EOD teams are tasked to support civil authorities, the EOD battalion will approve response communications plans to verify that teams will be able to maintain secure and uninterrupted communications, both voice and data, with local state, federal and tribal emergency services. The EOD battalion will also verify company response SOPs to ensure that primary and secondary communications systems are available for incident response.

INTELLIGENCE

3-6. In accordance with the Posse Comitatis Act and federal law, no intelligence collection will be conducted by EOD forces within the US. Battalion and company operations sections should work closely with the installation PMO protection cell to provide situational awareness for known criminal and terrorist threats in the assigned response area. Any photographic or material evidence collected in response to civil authorities will be turned over to them at the end of the incident.

CREW/COUNTERMEASURES

3-7. ECM use during an EOD response in the homeland involving suspected radio controlled IED (RCIED) must be coordinated through the Feral Bureau of Investigation (FBI) and in accordance with DOD Instruction 3025.1. The FBI is responsible for all domestic use of ECM for RCIED and has established procedures that facilitate ECM employment by military EOD when responding to suspected RCIED on and off military installations.

3-8. Before using ECM, the EOD on-scene commander will initiate the RCIED ECM protocols as established in the current FBI military EOD domestic ECM notification checklist by calling the FBI Strategic Information Operations Center or the local FBI Special Agent Bomb Technician. Before using ECM, the EOD on-scene commander will inform the incident commander so that they are aware of the potential impacts that RCIED ECM may have on incident operations.

COMPANY COMMAND SECTION

3-9. DOD Directive 3025.18, Defense Support of Civil Authorities and DOD Instruction 3025.1, Defense Support of Civilian Law Enforcement Agencies outlines the military's responsibilities to civil authorities. The EOD company command section will ensure that they have an understanding of these documents. Company commanders and first sergeants will educate their Soldiers on this directive to enable them to have a better understanding of the importance of civil support

3-10. The EOD company command section is responsible for coordinating all EOD support to local, state, federal, and tribal law enforcement agencies. This support includes emergency response to all explosive ordnance found on and off DOD installations, as well as planned support to the Drug Enforcement Agency (DEA), FBI, Department of Energy (DOE), Department of State (DOS), and the USSS.

3-11. Company SOP for homeland response requirements will be maintained and updated annually by the company command section. Every Soldier that will be responding to civil authorities will read and understand the company SOP when they report into their unit. Unit SOPs will include state and local regulatory requirements as outlined in battalion and group base orders.

3-12. The EOD group will ensure that all appropriate Memorandums of Understanding (MOU) and Memorandum of Agreement (MOA) are signed with appropriate state environmental agency and the sister service EOD units that will take responsibility for areas around their installation.

3-13. Both the MOU and the MOA should establish the responsibilities and capabilities of the closest EOD unit in order to have an understanding of what type of capability is needed based on the type of incident that is reported. Any time military explosive ordnance is found, a military EOD unit will be tasked to respond to the incident.

3-14. Technical assistance provided to local, state, federal, and tribal authorities is provided by the EOD company. This should also be included in the capabilities brief given by the command section. Technical assistance includes but is not limited to, interpreting X-rays and pictures that are provided by law enforcement in order to determine what type of threat they may have, advising and assisting on courses of action, and render-safe and/or disposal procedures in accordance with applicable MOU's and MOA's. The EOD company must know where available disposal sites are within their area of responsibility. In some instances the requesting agency will need to provide a safe disposal area. Establishing an MOA will allow the EOD team to access the closest disposal site on or off a DOD installation. The EOD team will ensure that they have the appropriate Environmental Protection Agency (EPA) permits prior to disposal in accordance with state EPA.

3-15. In order to respond in the homeland every EOD Soldier must be certified as a first responder. This certification is provided by the Federal Emergency Management Agency (FEMA) under the Army emergency management program, national information management system, incident command system, and rescue techniques for first responders (AR 525-27).

3-16. The command section will ensure compliance with Code of Federal Regulations (CFR) 29, which states that every emergency responder that may come in contact with a hazardous substance must have passed the Occupational Safety and Health Administration (OSHA) physical and met the operational requirements for Hazardous Waste Operations and Emergency Response (HAZWOPER) technician.

3-17. Other regulations, policies and directives that the command section must ensure they understand, and are in compliance with are CFR 40, Protection of the Environment, CFR 49, Transportation, Presidential Policy Directive (PPD) 17, Countering Improvised Explosive Devices, and PPD 7, Critical Infrastructure Identification, Prioritization, and Protection, and the Military Munitions Rule.

3-18. The DOE has several sensitive sites throughout the US. The EOD command section will ensure that they are aware of all sensitive sites to include any sensitive shipments that may be moving through their area of responsibility. The command section must be able to provide teams to an area that contains sensitive items in a timely manner.

COMPANY OPERATIONS SECTION

3-19. The operations section is responsible for planning the operations, organization, and training as directed by the commander. The operations section also writes operational directives and plans and orders, as well as training schedules. The operations section prepares courses of action and recommends actions or decisions to the commander for the accomplishment of the mission.

3-20. The company operations section will ensure that all VIPPSA support is manned in accordance with policy. VIPPSA support will often be requested on short notice. The operations section must be prepared to task EOD teams in order to support the mission. If the company cannot support the mission then the operations section will inform their parent unit that they are unable to support the mission as soon as possible.

PLANNING CONSIDERATIONS

3-21. Several considerations go into an EOD emergency response or planned operation. Proper planning will ensure that the EOD company provides the appropriate level of response for every request. Some of the considerations that must be considered are the type of incident or accident as well as the area that the incident or accident is located.

3-22. The EOD company will have contingency plans in place in the case that an incident or accident occurs. In order to ensure that the company is able to respond throughout their area of responsibility they must know the size of the response area and the capabilities of all DOD and civilian bomb disposal capabilities in the area.

3-23. If EOD support is imminent the company may need to strategically place teams throughout their area of responsibility. If the response area is large the company will coordinate with air assets, both fixed and rotary wing, so that they are able to respond to their entire area.

3-24. In the event of a large scale incident or accident the EOD company command section may integrate with the incident commander at the site in order to provide command and control to the EOD teams that are

Chapter 3

assisting in the response. The major incidents or accidents that may require the presence of the command section include natural disasters, terrorist incidents, mass casualty incident and accidents involving explosives.

3-25. The EOD company will have all appropriate tools and equipment to respond to any incident or accident. This will include EOD CBRN capabilities. All tools and equipment must be in working condition and able to deploy if needed.

3-26. While every EOD team has the capability to respond to all types of incidents or accidents involving explosive ordnance, there may be times when more specialized assets are needed. Army technical escort or civilian hazmat support may be needed in the case of some CBRN incidents or accidents. This determination can be made by the EOD team leader once on scene in accordance with unit SOP.

3-27. EOD teams must ensure that they are responding to law enforcement acting in an official capacity. Police escort allows the EOD team to arrive at the incident site in the fastest manner possible. Once on site the EOD team must have site security to ensure that if there is a high order detonation the area is clear. Medical and fire support must also be on site in the event that they are needed. In order to avoid confusion these requirements will be established by the EOD company and the local, state, federal, and tribal authorities before an incident or accident occurs.

3-28. The EOD team responding must know the appropriate route to take to get to the incident site in the fastest manner possible. Sue to the EOD transporting hazardous items, the EOD team leader is responsible for ensuring that all appropriate safety precautions and HAZMAT transportation requirements are in accordance with 49 CFR. The EOD team will coordinate with local first responders to ensure that the appropriate hazardous transportation route is known by both the EOD team and the police escort. In accordance with AR 75-15 and AR 190-11, EOD teams conducting emergency response are exempt from the requirement to carry weapons.

3-29. Depending on the type of incident there may be a need for protective works to preclude collateral damage at the incident site. If in an urban area or an area that cannot withstand a high order detonation then the EOD team may need to request a containment vessel, bomb blanket, or sandbags. Law enforcement will be made aware of the possibility of having to supply these items upon the request for EOD support.

PRE-PLANNED LAW ENFORCEMENT SUPPORT

3-30. The EOD company may be requested to conduct pre-planned law enforcement support to drug suppression teams, hostage rescue teams, counter terrorism, and special response teams. Agencies will request EOD support through the Office of the Secretary of Defense and the request will be sent down through the EOD companies' chain of command.

3-31. While these planned operations can be conducted without EOD support, if there is a possibility of WMD, or military ordnance an EOD team should be included in the mission.

3-32. There will be times when a government agency may need EOD to provide only technical support. If this type of support is requested it will be written into the request exactly what role EOD will play in the operation.

3-33. EOD will also conduct planned operations on military posts and DOD installations. This support will usually include supporting military police or federal agencies for planned missions that may include explosive ordnance.

Rehearsing

3-34. During the initial rehearsal stage, the EOD command or operations section will advise the supported agency on the best way to utilize EOD assets. Governmental agencies may not be fully aware of all of the capabilities that an Army EOD team can provide versus a civilian bomb disposal unit. The unit commander will ensure that supported agencies understand the limitations imposed on federal military forces in accordance with DOD Directive 3025.18.

3-35. EOD is usually on standby during a planned operation and will be pre staged in order to quickly respond when they are called in. If the supported agency is conducting rehearsals prior to the operation and would like EOD to participate they must allow adequate time for EOD team to prepare.

3-36. When there are time restraints, and a rehearsal is not possible, the EOD team should conduct a talk through with the supported agency versus a walkthrough. This will allow the EOD team to know where they should stage and how to find out if they are needed.

Execution

3-37. Execution of the operation is based on the requirements of the supported agency. EOD will advise on proper actions to be taken if an explosive ordnance/IED/homemade explosive/WMD is located. When the EOD team is called in to mitigate the explosive ordnance/IED/homemade explosive/WMD they must do so in a safe and efficient manner so that they do not delay the supported force.

After Action Review

3-38. At the completion of a planned operation an after action review (AAR) may be conducted. The AAR will cover all aspects of the EOD support provided. The EOD team may also recommend changes that need to be made if a similar operation is conducted again. Since EOD is the supporting unit during a planned operation then the AAR will be conducted by the supported agency. After action reviews are usually conducted for high profile incidents.

RESPONSE OPERATIONS

3-39. EOD teams are on standby throughout the homeland to respond to local, state, federal, and tribal law enforcement 24 hours a day. All EOD requests for assistance inside the US and its territories should be coordinated through local law enforcement.

3-40. A primary team will be on call ready to respond in each designated response area if a call is received. The duty team may consist of two to three EOD Soldiers, with a team leader certified EOD Soldier as the duty officer. The duty team will have a duty phone so that they can be reached at any time.

3-41. There will also be a secondary team on standby if there is a need for additional teams. The secondary team will be deployed at the discretion of the command section depending on the number and types of EOD support requests received.

3-42. During regular duty hours the team will be prepared to leave the base within 30 minutes of receipt of the call. If an incident request is received during non-working hours the team will leave the base within 60 minutes of receipt of the call.

3-43. If the EOD company receives a direct request from civil authorities they will inform their parent battalion. This allows the battalion to keep an over watch of what type, and the amount of incidents, that their company's teams are responding to. The EOD team will also inform the base emergency operations center with a brief description of the incident report.

3-44. Unit SOP will dictate further actions for the EOD team responding to incidents or accidents within the homeland. The command section will ensure that the unit SOP is current with up to date contact information and any other information that the teams may need to know when conducting homeland response.

3-45. In the event that EOD support is requested for several incidents within the same geographical area the company command or operations section will prioritize the missions. This prioritization will be determined on what is deemed to be the most serious incident, location of the incident and impact to the community.

3-46. If the operations or command section determines that there are several serious incidents that require immediate response, the command or operations section may decide to deploy additional teams in accordance with unit SOP. If the EOD company continues to receive more high priority missions then the EOD company will continue to deploy all of their assets until they have been exhausted.

3-47. Additional assets may be requested by the EOD company if they are unable to respond to all high priority incidents. Additional support may be provided by both military and civilian bomb disposal.

3-48. Civil authorities may report an incident and assign it a high priority. If there is a delay in response time, the company will inform the civil authorities that there will be a delay due to higher priority mission load, availability of assets, or transportation issues.

Incident Reporting

3-49. Department of the Army (DA) Form 3265 (Explosive Ordnance Incident Report) must be filled out for every incident that EOD responds to. In addition to the DA Form 3265, EOD teams will use Explosive Ordnance Disposal Incident Management System (EODIMS) database for incident record keeping. The DA Form 3265 and EODIMS record are reports that give a general description and timeline of the actions that the EOD team took throughout the response. The classification of these depends on the information that it contains. The EOD group commander will determine what information must be put into the report.

3-50. The DA Form 3265 and EODIMS must be completed and sent to the parent EOD group through the EOD team's chain of command within 24-48 hours of completion depending on the category of incident. If there is a need for a further narrative, then an additional page can be attached to the DA Form 3265. If any explosives or energetics are used during the response, the team must ensure quantities are annotated on the DA Form 3265 and in the EODIMS Record.

3-51. When developing the EOD report it is important to use approved terminology that can be easily understood by all. The Weapons Technical Intelligence IED Lexicon is intended to encompass the broad spectrum of IED employment scenarios, the variety of IED devices, and their critical components. The lexicon was developed and approved by subject matter experts from military and civilian agencies as well as North Atlantic Treaty Organization.

3-52. For large scale events that are considered to be unusual or serious a storyboard may be completed in accordance with unit SOP. A storyboard allows the team to expand on the information that was gathered throughout the incident or accident. The storyboard may also contain unclassified pictures and X-rays in order to show what the EOD team responded to. For more information refer to AR 190-45, Law Enforcement Reporting.

3-53. If the civil authorities determine that evidence gathered at the scene contains explosives and must be kept for evidence, a DA Form 4137 (Evidence/Property Custody Document) or the equivalent civil law enforcement custody report, must be filled out and retained by the EOD team. When retaining evidence, the EOD team should consult with the servicing judge advocate or legal advisor as soon as practicable to prepare for litigation. The EOD team will take the evidence and store it in the explosives bunker with the report until civil authorities request it for use in the criminal case, or request that it be destroyed. Storage for other than emergency needs must be coordinated through the installation.

3-54. The agency that EOD provided support to may request a copy of the DA Form 3265 and EODIMS report that was filled out for the response. This form may be used in court as evidence. If the agency requests the form or other information arising from the response, the company commander must consult with the servicing judge advocate or legal advisor immediately to determine whether it is releasable.

EOD TEAM RESPONSE

3-55. The duty team will be able to respond to any request for assistance. Assistance may be requested directly from local, state, federal, or tribal authorities. Once approved by the commander, the team will respond to the request.

3-56. Request for EOD assistance to respond to incidents or accidents on a DOD installation may come directly from authorities at the installation. These responses may include suspicious packages or UXO on ranges.

3-57. When a team leader receives a call they will immediately notify the duty officer. The duty officer will then notify the command section and higher headquarters with all information that they have gathered in regard to the type of request. The team leader must try to gain as much information about the type of incident that they are responding to from the person calling for support.

3-58. Unit SOP will dictate the type of information that must be gained before responding. The information that must be collected is point of contact information and phone number, or police channel, description, and location from the caller, as well as where to link up with the incident commander, and if the area is secured.

3-59. AR 190-11, Ch 7-1.b, states that EOD teams responding to off-station accidents or incidents will transport necessary explosive ingredients in accordance with requirement established by the senior mission commander concerned.

3-60. C-4, TNT or military dynamite are considered to be category II explosives. According to DOD 5100.76, Physical Security of Sensitive Conventional Arms, Ammunition, and Explosives, category II explosive used for operational purposes may be secured in a combat vehicle, trailer, or other configuration required by operational requirements.

3-61. Keeping an accurate mission timeline allows the EOD company to keep track of the time requirements it takes to handle a call. A timeline may also be important in a criminal case. Once the call is received the mission timeline will begin.

CONSIDERATIONS

3-62. Information gained from the request for support can be used to formulate a plan before arriving on scene. Considerations based on that information will determine what the EOD team will take when responding.

3-63. Important considerations will be the type of incident as well as the location. Responding in the homeland requires that the EOD team have as little effect on the civilian populace as well as the infrastructure as possible. Because of this the tools and equipment must be used in a way to cause as little damage as possible.

3-64. If the EOD team determines that due to the type and location of the incident additional assets are needed then they should request those assets through the incident commander. Additional assets can aid in providing protective works or providing a larger cordon of the area.

3-65. Proper communication with local, state, federal, or tribal authorities as well as higher headquarters should be established before leaving base. Cellular phones can be used as a primary communication source but may not always be available depending on conditions. During incidents on post in support of installation, teams should maintain FM communications with Range Control and ensure all MEDEVAC frequencies are loaded.

EOD TEAM RESPONSIBILITIES ON SCENE

3-66. Throughout the response the team leader is responsible for ensuring that the actions of the EOD team do not endanger personnel and property, to include civilians, law enforcement, military, and the EOD team. In order to provide this protection the team leader will ensure that there is a proper cordon set around the hazardous item. If the team leader determines that the cordon should be expanded they should immediately inform the incident commander. Make sure that the incident commander understands that the item could detonate at any time throughout the call.

3-67. Maintaining contact with the incident commander as well as company headquarters throughout the call is essential. Before any action is taken the team leader needs to inform the incident commander and give the incident commander time to inform their personnel.

3-68. When applicable X-rays will be taken of the explosive ordnance/IED/homemade explosive/WMD. These X-rays will allow the team leader to be able to make the decision of what type of action to take in order to mitigate the threat. The X-ray may also show whether the item is a credible threat.

3-69. If the item is determined to be an IED then the EOD team will perform actions based on unit SOP. After the IED has been rendered safe, the EOD team will search the area for additional threats.

3-70. The team leader or duty officer will coordinate with the incident commander on the gathering of evidence from the device. Once the team has determined that the area around the device is safe to approach then the incident commander may send in their own personnel to gather the evidence.

3-71. In the case of a UXO there may be a need for proper protective works to be set around the ordnance. The UXO may be in an unsafe condition and disposal on site may not be possible. At this time the team leader will discern whether to perform a RSP on the UXO so that they can transport it to a safe disposal area.

3-72. If there are differences in opinion on what type of actions should be taken on scene between the EOD team and the incident commander refer to unit SOP. Unit SOP dictates procedures that can take place on scene. If differences between the EOD team and the incident commander cannot be resolved then the EOD team needs to immediately report the situation to their company commander, at this time the EOD team will support in an advisory role.

SCENE TURNOVER

3-73. At the completion of the call inform the incident commander that the threat that was reported, as well as any other threats located during the call, has been mitigated. Conduct a back brief with the incident commander in order to show exactly what the threat was. This will allow the incident commanders report to mirror the EOD teams report.

WARNING

Never inform the incident commander that the area is free of all explosive hazards. There may be additional hazards in the area that were not found.

3-74. The team leader may also conduct an AAR with the incident commander which will cover what went right with the call and what could have been done better. This will be conducted in a professional manner so that the advice that is offered will be useful to the incident commander.

3-75. If there is evidence that has been collected that contains explosives or explosive remnants the incident commander may ask the EOD team to take control of the evidence for safe storage. The EOD team leader will confirm with the duty officer and company commander that the evidence may be accepted into military custody and stored. The team leader must receive the proper documentation from law enforcement. A chain of custody receipt must be maintained at this time until the EOD team is directed to either turn over the evidence or destroy it. Before leaving the site contact the duty officer or company command in accordance with unit SOP.

VERY IMPORTANT PERSONS PROTECTION SUPPORT ACTIVITY

3-76. As directed by AR 75-15, EOD personnel are provided to the USSS, the DOS, and the DOJ upon request. The procedures herein address USSS missions because they are most frequent. However, procedures are generally the same for the other organizations.

3-77. The primary responsibility of the EOD personnel supporting the USSS is to detect explosive hazards, and advise the USSS so that the principal may be evacuated. The primary responsibility for a device, if one is found, lies with the law enforcement agency with jurisdiction for the area. If the principle is visiting a military installation, military EOD will conduct the necessary procedures.

MISSION RECEIPT

3-78. VIP support-mission assignments are passed to EOD companies by the EOD group or battalion. Upon receipt of the assignment, the senior EOD Soldier will contact the special agent in charge of the missions for any special information or instructions. This will include the required arrival time for the team, where to meet the special agent in charge or the Technical Security Division representative, and what special equipment and clothing are required.

PRE-MISSION PREPARATION

3-79. Prior to the receipt of any VIP-support mission, the EOD company commander will ensure the teams have the items required. All teams must be familiar with the guidance outlined in the Hazardous Device Countermeasures Manual, published and furnished by the USSS. The senior EOD Soldier and team members will prepare for a mission by reviewing AR 75-15, the Hazardous Device Countermeasures Manual, as well as local policies.

MISSION EXECUTION

3-80. Upon arrival at the designated location, the senior EOD Soldier reports to the USSS representative. The USSS should provide the tentative mission schedule, the areas to be searched, and the route and itinerary to be followed by the dignitary, including alternate routes. Information should also be provided on the location and room numbers of the security room, the package holding area, and the communications room; the EOD capabilities of local authorities; and the threat summary.

3-81. Before and during the dignitary's visit, the senior EOD team surveys all areas to be covered and plans for the search. Prior to the search, the senior EOD Soldier coordinates with technical security division or USSS agent for the use of military working dogs or uniformed division search dogs. The senior EOD Soldier recommends any requirements or equipment needed to support the EOD teams.

3-82. Normally, the responsible USSS agent plans the search and designates the explosives holding area and evacuation routes. This agent is a communications link for the EOD team. However, in the absence of a Technical Security Division agent or other designated representative, the senior EOD Soldier may have to do those duties. The responsible agent relies on the senior EOD Soldier for technical advice and recommendations about explosives. Therefore, the senior EOD Soldier must carefully consider recommendations before giving them to the responsible agent. The recommendations will be used in the coordination and direction of the actual search.

3-83. Before the search, the senior EOD Soldier meets with the USSS agent and local law enforcement officials. At the meeting, they determine the available disposal facilities and local bomb squad capabilities and responsibilities. They also determine the evacuation routes between the visit location and disposal facilities. When established disposal facilities are not available, they select a temporary site and select routes to the disposal sites that avoid crowds and main traffic areas. They also arrange for an escort vehicle in case it is necessary to move an explosive device. Explosives holding areas must be as far away as practical from the areas the VIP is scheduled to visit.

3-84. Search procedures for facilities and adjacent areas are outlined in the Hazardous Device Countermeasures Manual. Security of search and post search areas is the responsibility of the USSS. EOD personnel will not be used to secure areas or stand post.

3-85. EOD personnel advise the USSS agent when the search is completed. The advance agent then determines where EOD personnel are to wait. The senior EOD Soldier furnishes technical advice for identifying an explosive or incendiary device. The senior EOD Soldier also notifies the agent of the findings. If necessary, this agent helps carry out the evacuation plan.

3-86. When the principal arrives, EOD personnel may, at the discretion of the responsible USSS agent, be assigned to that agent during the visit to provide immediate on-the-scene technical advice. If an explosive item is found, EOD personnel may be charged with helping the responsible agent start evacuation procedures. During the actual visit, any extra EOD personnel may be put on standby status.

3-87. If a bomb, incendiary device, or any suspected item is found during a search, it must be reported immediately to the USSS. Even though speed is critical in reporting a situation of this type, care and discretion must be used in reporting such an incident. Under no circumstances should the find be reported to anyone other than a representative of the USSS.

3-88. During the performance of the mission, every effort must be made to divert public attention from EOD or military working dog activities. EOD should immediately refer any inquiries about EOD support activities to the responsible USSS representative.

POST MISSION ACTIONS

3-89. When the mission is completed, the senior EOD Soldier has an exit briefing with the agent in charge. The purpose is to discuss any problems that came up during the mission. Any unresolved problems are immediately reported to the EOD company commander. Just before the EOD team departs the site, the senior EOD Soldier contacts one of the following to determine if any missions are pending: the USSS coordinating center (when in operation), the EOD company, or the senior EOD Soldiers home unit.

Chapter 4
Support to Special Operations Forces

The range of military operations describes a need for ARSOF in joint, combined, and multinational formations for a variety of missions—from humanitarian assistance to major combat operations, including conflicts involving the potential use of WMD (ADRP 3-05). With the IED being used as the weapon of choice by organizations around the world, and being designated as an enduring threat, the need for EOD support is greater than ever. Army EOD is the service common solution to ARSOF and must be capable of supporting all required SOF missions.

OPERATIONAL ENVIRONMENTS

4-1. Special Operations Forces are deployed throughout the world in support of GCC and ambassadors. Most of these locations will be in non-permissive or semi-permissive environments that lack any US conventional forces. These locations may require a small force that has the ability to conduct high risk, discrete operations, or a force that is tailored to provide specific training to indigenous forces.

4-2. Large scale conventional operations are also heavily supported by ARSOF. These operations may include the targeting of individuals that pose a threat to US and multinational forces or hostage rescue, as seen during offensive operations in recent combat operations. ARSOF will be subjected to all types of explosive threats during operations and must be prepared to deal with these threats as they are encountered.

CORE ACTIVITIES AND OPERATIONS

4-3. The EOD and SOF command relationship is METT-TC dependent. The EOD team and platoon should fully integrate with supported Operational Detachment Alpha (ODA)/Operational Detachment Bravo (ODB)/Ranger formation while providing frequent SITREPs to the EOD company.

4-4. EOD provides a unique set of capabilities not organic within ARSOF formations. These capabilities include counter-IED, advanced IED circuitry recognition, sensitive site exploitation, homemade explosive and technical intelligence subject matter expertise as well as the ability to render safe and dispose of all explosive ordnance/IED/homemade explosives/WMD.

4-5. The subject matter expertise provided by EOD is utilized by ARSOF throughout their core activities and operations. Core activities are operationally significant, unique capabilities that SOF apply in different combinations tailored for an operational problem set. Core operations are the military missions for which SOF have unique modes of employment, tactical techniques, equipment, and training to orchestrate effects, often in concert with conventional forces (ADRP 3-05).

4-6. There is no difference in the core competencies provided by EOD to ARSOF as any other conventional force. Where this support differs is in the ability for the EOD Soldier to integrate into the ODA/ODB/Ranger formation. This integration includes special mobility skills and maturity which allow the EOD Soldier to move and communicate in a similar manner to the ODA/ODB/Ranger formation and make the correct decisions in regards to explosive threats that are encountered. It is imperative that EOD Soldiers have a complete understanding of the operations, activities, and terminology of special operations in order to make these decisions.

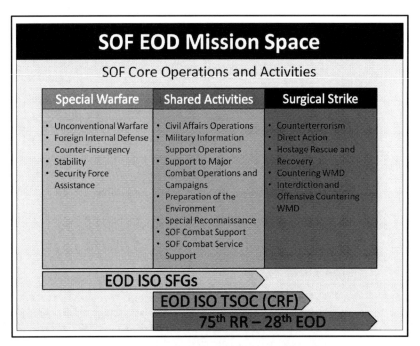

Figure 4-1. SOF EOD Mission Space

EQUIPMENT

4-7. ARSOF conduct operations in all environments. EOD tools and equipment must be configured to allow the EOD teams to integrate into the supported unit and operate in as safe a manner as possible. Due to the nature of the missions that the EOD teams will support, lighter and compatible equipment that still allows for safe operations is essential. Due to the heavy usage, these items will need to be replaced at a higher rate. Training as you fight is a critical component to success in the OE. EOD companies in support of ARSOF must be provided the same tools and equipment during training in order to gain confidence and proficiency. The list below is a basic layout of what each Soldier will use on SOF support operations

- Assault pack
- Plate carrier
- Compatible dismounted communication equipment
- Compatible Night vision
- Lightweight Metal Detector
- Lightweight Individual Hook and Line kit
- Lightweight disruptor
- Lightweight robots

TRAINING

4-8. Mission success relies upon the quality of pre-deployment training conducted with the ODA/ODB/Ranger formation that the EOD team will deploy with. During the training, the EOD team will have the opportunity to gain trust and confidence with the supported unit. It is also essential during this time to educate the ODA/ODB/Ranger formation on the capabilities, constraints and limitations of the EOD team while ensuring that the limitations cannot be mitigated by mobility training and equipment substitution/modification.

4-9. The requirement for EOD operations within an ODA/ODB/Ranger formation is primarily dismounted; therefore training plans must be adjusted accordingly. The EOD company needs to focus on the team's ability to support combat operations with a smaller amount of equipment and support. This training will also focus heavily on critical risk management. Each EOD technician needs to have the ability to make difficult, time critical decisions based on the mission and their own capabilities.

4-10. AROSF may use explosives that are not readily available in conventional deployments. EOD teams must become proficient in the use of these explosives during pre-deployment training. Some explosives may not be accessible until the EOD team is in theater. The EOD team must educate themselves on all explosives that may be used. Cross training with the ODA/ODB (18C) Engineers will allow for interoperability during deployment.

4-11. Sensitive site exploitation training must be conducted, as most areas of the ODA/ODB/Ranger formation missions are specifically targeting these sites. EOD is expected to understand what forensic evidence is important and how it applies to enemy TTP's and/or operations. The EOD team must have the ability to be trained on and to conduct exploitation in accordance with supported ODA/ODB/Ranger formation SOP.

4-12. Because ARSOF deploys throughout the world, a counter-IED reach back capability may not be readily available. Without this reach back ability the EOD teams will be required to have an advanced understanding of IED circuitry and functioning. Pre deployment training should focus on providing advanced electronics to all EOD technicians in order to provide information from targeted sites.

4-13. CREW/ECM systems used by the ODA/ODB/Ranger formation must be incorporated into the training. All EOD technicians must have a complete understanding of the capabilities of these systems. The EOD team should ensure that all of the tools and equipment work while the CREW/ECM is on.

Special Forces Pre Mission Training

4-14. ARSOF goes through pre-deployment training similarly to a conventional company's road to war. The timeline for the training is approximately eight months and has gates that resemble the ARFORGEN process. The tasks are trained individually, collectively, and then the ODA/ODB/Ranger formation is validated through a culminating event.

4-15. During the eight month training cycle, EOD teams are afforded multiple training events that are absolutely critical to the success of this enduring mission. The training plan for an ODA/ODB/Ranger formation focuses on mobility, security, targeting operations, and Find, Fix, Finish, Exploit, Analyze, and Disseminate (F3EAD). The training process allows the EOD team to integrate into this training and increase the individual Soldier's skills within both the EOD team and the ODA/ODB/Ranger formation as a functioning support member.

4-16. The success of the EOD team's ability to support the ODA/ODB/Ranger formation depends heavily on their ability to integrate into the supported unit during missions. Below is a list of Special Forces specific training that should be completed by the EOD team to ensure mission success.
- Marksmanship at all levels from urban to sniper training.
- Medical certifications higher than combat lifesaver and focused on traumatic injuries.
- Maneuver training on multiple platforms used in theater (horseback/off road motorcycles/quads/helicopter/boats).
- Cross training with the other members of the ODA/ODB/Ranger formation on their tasks.
- Communications training on systems used by ARSOF.
- Intelligence training related to EOD operations and input into the targeting structure.
- Mounted and dismounted movements using night vision.
- Immersion language and culture training.
- Urban assault breaching.

Deployment/Re-deployment

4-17. Deployment and Re-deployment are higher priority for the EOD platoons that support the ODA/ODB/Ranger formation. The availability of airframes and moving of equipment for ARSOF make the EOD platoon movement simple. The ODA/ODB/Ranger formation must have the platoon equipment and personnel integrated into their movement plans early.

4-18. Deployment requirements become a cyclic process. As one platoon prepares to deploy and they complete those activities, another is returning, requiring redeployment activities to begin. EOD companies support the ODA/ODB/Ranger formation through the request for forces process; therefore they are not under the control of the supported unit. Because of this, all of the individual deployment requirements for Soldiers, which would normally be completed by the EOD battalion, fall on the EOD company with battalion support.

4-19. Due to the physical location of most of the ODA/ODB/Ranger formation, moving an entire EOD team's worth of equipment is not possible. If deploying to a mature theater, the command and control element in theater must manage a small supply stock to support the EOD teams with the supported unit. This includes Class V and EOD specific equipment.

4-20. Monitoring of the team's requirements and supply levels is critical. Movement into some of these locations can take weeks due to lack of aircraft support and physical location of the ODA/ODB/Ranger formation. The quick method of delivering equipment is airdropped. This however, increases the risk of damage and loss of sensitive equipment. The airdrop method also keeps the command and control element from physically checking with the teams.

Reporting

4-21. EOD teams will be required to submit reports in the same manner as they would in support of any other organization. ARSOF may require additional information from the EOD teams for their reporting systems. Reports submitted in support of SOF may have classifications that exceed routine EOD reporting. These reports may be censored by the supported unit before submission. Information provided by EOD enable increased mission planning for the ODA/ODB/Ranger formation operations and increases survivability for all elements involved. Additional reporting requirements should be discussed during pre-deployment training.

75TH RANGER REGIMENT (AIRBORNE)

4-22. Rangers are a rapidly deployable airborne light infantry organized and trained to conduct highly complex joint direct action operations in coordination with or in support of other special operations units of all Services (JP 3-05). EOD support to the Ranger Regiment is provided by the 28th EOD(A) Company. The 28th EOD(A) is the only Army EOD airborne company and is deployed in direct support of the 75th Ranger Regiment.

SPECIAL FORCES GROUPS (AIRBORNE)

4-23. ODA's have an increased likelihood of being targeted by the enemy due to the nature of their operations. The ODA's have a mission that specifically requires freedom of movement in and around the local populace on a constant basis. This does not allow for variations in travel or times.

4-24. EOD teams will be fully integrated into the supported ODA. This integration will sometimes force the EOD team to conduct operations in a manner that is considered more hazardous due to time constraints and limited access to tools and equipment. In order to prepare for every contingency possible the EOD team needs to conduct a thorough pre mission plan. Items that must be considered are:

- Route reconnaissance.
- Historic IED activity.
- Assault force threat brief.
- Communications/EOD key calls.
- Grid reference guides.

4-25. Due to the physical location of the operations that the ODA's conduct the EOD teams will not always be prepared to deal with every threat that is encountered. The EOD team should prepare equipment drops that are specifically tailored to possible threats that may be encountered. If the EOD team is forced to deal with a threat that they are not properly equipped to deal with the tailored equipment can be flown into them. The following is a list of possible missions that may require a tailored equipment drop:

Support to Special Operations Forces

- CBRN operations.
- Multiple UXO.
- IED operations.

4-26. The most important thing for the EOD team is to have a complete understanding of their capabilities, constraints, and limitations. The operation that is being conducted may have an increased risk due to the conditions. If the EOD team is not properly prepared to neutralize the threat that is encountered they may need to mark and bypass. The EOD team must keep an accurate record of all items that have been marked.

Chapter 5

Chemical, Biological, Radiological, and Nuclear Operations

Weapons of mass destruction are CBRN weapons capable of a high order of destruction or causing mass casualties and exclude the means of transporting or propelling the weapon where such means is a separable and divisible part from the weapon. (JP 1-02) ADP 3-0, Unified Land Operations, states that one of the most challenging potential enemies comes in the form of a nuclear-capable nation-state partnered with one or more non-state actors through ideological, religious, political, or other ties. All EOD technicians are trained to provide first response to suspected WMD. They act in coordination with more specialized national WMD response assets.

CHEMICAL, BIOLOGICAL, RADIOLOGICAL AND NUCLEAR THREATS AND HAZARDS RESPONSE

5-1. AR 75-15 states that any time an EOD unit responds to an incident involving actual or suspected CBRN devices and or materials the unit will immediately notify the Army Operations Center (DAMO-OD-AOC) through command channels. A CBRN threat is the intentional employment of, or intent to employ, weapons or improvised devices to produce CBRN hazards (FM 3-11). CBRN hazards include toxic industrial materials (TIM). TIM includes toxic industrial chemicals (TIC), toxic industrial biological (TIB), and toxic industrial radiological and in sufficient quantities, may pose a danger to the environment, people, and animals. EOD forces have the capability to respond to CBRN threats and hazards whether they are an intentionally release of CBRN material or an unintentional release of toxic industrial materials.

RESPOND TO CHEMICAL AND BIOLOGICAL EXPLOSIVE ORDNANCE INCIDENTS AND ACCIDENTS

5-2. Upon arrival at the incident or accident site, EOD will position an EOD command post to coordinate EOD and supporting unit operations. The command post supervisor will coordinate with the designated incident commander or CBRN officer for updated information and mission requirements. The EOD command post provides information on the expected type and extent of contamination. In an emergency, an EOD company can decontaminate EOD personnel and equipment only. CBRN forces must supplement EOD personnel as soon as possible to complete the mission. The incident commander or supported agency must provide all decontamination, resupply, medical, and security support.

5-3. Disposition of the ordnance is based on information provided by EOD and on the current situation. Chemical and biological explosive ordnance of intelligence value must be rendered safe and removed for exploitation by the appropriate agencies. The incident/mission commander must coordinate with required supporting elements to secure the items and transport them to the designated area for release to the CBRNE Response Team.

OVERALL RESPONSIBILITIES

5-4. During combat operations the locating unit will send an EOD 9 Line Report through their chain of command requesting EOD support. The EOD 9 Line Report may or may not provide information on chemical or biological hazard detection. The EOD team should be aware of contamination indicators prior to arriving on scene.

5-5. In the homeland, the requesting unit or agency will provide the location of the incident, identification of the explosive ordnance items, chemical and biological hazard survey information, and the name and location of the individual authorized to grant access to the incident site. It must also identify security and evacuation measures being enforced.

5-6. When the EOD Company is assigned a chemical incident or accident, it sends a team to the scene by the fastest means. The team notifies the EOD battalion and coordinates with the chemical accident or incident response and assistance (CAIRA) officer. In some cases, the EOD team leader may be the first military representative on-site and will assume control. The EOD team leader is responsible for the safety of EOD personnel when they enter the hazard area.

5-7. Either the incident commander or the CAIRA officer is responsible for the management of chemical incident operations. The on-scene commander may assume all responsibilities of a CAIRA officer during operations in a hostile or uncertain environment (IAW AR 50-6). Resources available to the CAIRA officer for supporting operations include the following:
- Communications equipment (civilian and military).
- CBRN teams (recon and surveillance and decontamination platoons).
- Civil Support Teams (CST National Guard)
- Transportation (including air).
- Medical teams and facilities.
- Engineer personnel and equipment.
- EOD personnel.
- Security forces (military and civilian).
- Military police.

5-8. The EOD team that responds will handle the incident based on the resources available, the current situation, directives from the headquarters having operational control of the EOD Company, quantity and type of weapons, and types of agents. The team may be able to complete the incident without additional support from the EOD Company.

5-9. If the EOD team leader determines that the situation warrants more capabilities than the EOD team has, then the team leader will request additional assistance through the chain of command. The team leader must determine if the team can handle the incident by asking:
- How much contamination is present and how likely is gross contamination?
- How many ordnance items are involved and what are their fuze conditions?
- What procedures will be used on the ordnance?
- Can the supported unit help the team with additional personnel or equipment?
- Disposition instructions on items in question before transport?

5-10. If additional support is requested, the EOD officer will establish a command post in order to provide mission command of the incident site. The area may be approached from any direction if it is free of contamination. Suspected contaminated areas must be approached from upwind, and team members must don protective equipment prior to entering the area.

5-11. The area should be monitored with two forms of monitoring equipment if contamination is suspected. Surveys should begin at least 1 mile from the proposed command post site.. The team must take samples of suspect contamination during the approach to be sure the area upwind of the command post site is contamination free. If contamination is found at the proposed site, personnel must decontaminate themselves and select another site.

5-12. The EOD officer will:
- Designate a recorder.
- Assign an equipment specialist.
- Designate at least two EOD personnel to enter the contamination area. The EOD personnel will include a team leader and EOD specialist. Other personnel will be to provide "buddy aid" for casualties.
- Designate work party members.

- Assign a staff sergeant or above to be the emergency personnel decontamination station (EPDS) supervisor and two EOD specialists to assist.
- Ensure proper communication systems are used.
- Determine individual protective equipment requirements.
- Determine the need for and implement (if necessary) electromagnetic radiation (EMR) hazard precautions. The CAIRA officer or on-scene commander must be notified of any EMR hazards and notify supporting elements.

BEFORE RESPONSE

5-13. Response to the incident involves several planning requirements. These include gathering information, planning the route, and selecting the personnel, equipment, and procedures for the operation.

5-14. The EOD team will carry, as a minimum, detection equipment, first aid materials, a means of communication, EOD tools and equipment, and limited amounts of decontaminants.

5-15. The size and number of the munitions involved will determine equipment requirements. If total agent disposal cannot be done on site, the items must be packaged suitably for transportation and properly stored until final disposal determination.

5-16. Planning considerations for the EOD team leader include individual protective equipment requirements and type and amount of contamination. If the type of agent is unknown, the highest level of personal protective equipment must be implemented.

5-17. EOD Teams should use their service-approved personnel protective clothing ensembles. Although protective clothing is designed to protect against penetration of chemical agents, some penetration may occur. For this reason, any liquid contamination on the clothing should be removed and decontaminated as soon as possible.

COMMAND POST SETUP AND PERSONNEL RESPONSIBILITIES

5-18. The proposed site for the command post will be checked for possible contamination. There must be enough checks made to convince the EOD team leader that the entire command post area is free of contamination. The M256A1 detects and identifies lewisite and nerve, blister, and blood agents and is used to support unmasking procedures. The test takes 15 minutes to complete, and a color change indicates harmful chemical-agent vapor concentrations. The area will not be considered free of biological contamination unless intelligence information or physical evidence confirms it. If biological agent contamination is suspected, take samples and send them to a laboratory for identification. Unmasking procedures are outlined in FM 3-11.4.

5-19. When selecting a site for the command post, you must consider the mode of transportation to be used to get to the incident site. You will also have to establish an exclusion area as for conventional ordnance, with two exceptions. When there are no explosive components, you can reduce the exclusion area to a 50-meter radius. This is done with concurrence of the CAIRA officer or the on-scene commander. When the team has not yet determined what types or numbers of explosive components there are, the minimum exclusion area must have a 381-meter fragmentation radius plus a 50-meter EPDS distance.

5-20. A downwind hazard area must be established 2000 Meters (1.24 Miles) downwind from the accident/incident site. All unprotected personnel should be evacuated from the hazard area or required to wear chemical protective clothing. The on-scene commander will ensure the personnel conducting the evacuation also wear protective clothing.

Chapter 5

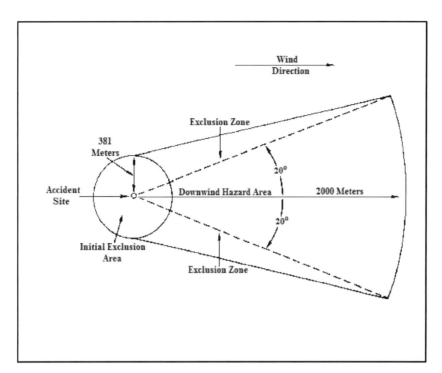

Figure 5-1. Example downwind hazard prediction area

5-21. The command post site must be upwind of the incident site or suspected contaminated area and be close to, but separate from, support element operations. No other element should be established closer to the incident site than the EOD command post.

Command Post Supervisor

5-22. The command post supervisor organizes the command post for on-site operations. The supervisor maintains contact with the headquarters having operational control of the EOD Company as well as the company HQ or EOD battalion. In addition, the command post supervisor is responsible for the following:
- Keeping the CAIRA officer or incident commander informed of the incident progress.
- Advising the commander on workparty procedures and operations.
- Checking the workparty methods and advising them on alternative courses of action.
- Directing research in support of the team.
- Monitoring log entries.
- Coordinating for external support.
- Inspecting the workparty to ensure proper dress out and equipment setup.
- Adjusting the exclusion area and downwind hazard area as new information is received and making recommendations to the CAIRA officer or on-scene commander. The supervisor posts this information to the situation map.
- Requesting disposition instructions for the contaminated waste, classified components, and other contaminated items.
- Notifying the survey team and decontamination team of the type and amount of chemical/biological contamination.
- Protecting classified information in the command post and making sure all classified materials are accounted for after the incident is completed.
- Preparing required reports and closing out the command post as required.
- Supervising the recorder, equipment specialist, and EPDS supervisor.

- Requesting and coordinating support needed for such things as protective measures, evacuation, medical, decontamination, and security.

Recorder

5-23. The recorder keeps a complete, accurate record, relays information received to the command post supervisor, and ensures that classified paperwork generated in the command post is properly marked and safeguarded.

Equipment Specialist

5-24. The equipment specialist is in charge of tool assembly as well as the preoperational and operational checks of equipment. The equipment specialist will also assist the workparty in putting on protective clothing and starting equipment. They will keep a log of all equipment that is missing, broken, depleted, or destroyed. Finally, they load the tools and equipment for return to the unit.

EOD TEAM RESPONSIBILITIES

5-25. Safety measures are the foremost responsibility of EOD personnel because of the lethality of chemical and biological hazards. Safety measures include the two-person rule, the exclusion area, the downwind hazard area, and contamination control. The two-person rule applies while any work is done on the ordnance. Chapter 1-8 of AR 50-6 provides details on this technique.

5-26. The EOD personnel must determine the location, number, and condition of the ordnance involved. They must also determine the fuzing and its condition, the presence or absence of agent leakage, and the weather conditions at the site. If rescue operations are required, or there is a serious public health hazard, the team must quickly evaluate the situation and take emergency measures. Performing the RSP, contamination control, and decontamination are the three main functions of the workparty. If the situation dictates, combine the EOD team functions and workparty functions. This is only advisable when the exact situation is known.

During Response

5-27. The EOD team should approach by the most direct route, keeping upwind and avoiding heavy vegetation if possible. The team should look for obvious contamination, such as agent deposits (liquid or powder) in the area and people or animals showing symptoms of agent exposure. If possible, chemical monitoring should be accomplished using two forms of monitoring equipment (i.e. improved chemical agent monitor, joint chemical agent detector and M8 paper).

5-28. When the team is close to the incident site, the team leader should survey the area and situation and decide where best to place the safety observer. The safety observer must be able to watch the other team members and operate the communications equipment.

5-29. The team will set up and maintain constant communication with the command post. The team also observes fuze precautions for the munitions involved and report all activities to the command post as they occur. They will only pass classified information by secure means.

5-30. The EOD team must determine the location, number, and condition of the ordnance involved. They will also determine the fuzing and its condition. The team will observe the presence or absence of agent leakage, and the weather conditions at the site.

Work party responsibilities

5-31. When possible, the work party will complete the RSP for the explosive components before decontamination starts. This will prevent the possibility of a detonation in case of accidental movement. In some cases, the fuze and fuze condition may allow some decontamination while preparing for the RSP.

5-32. The EOD team leader decides the safest method for the situation. In some cases it may be possible to detonate the UXO in place (for example, a heavily contaminated area or an isolated area). In all cases, the EOD team leader must have concurrence with the on-scene commander. **In-place detonation is not a**

Chapter 5

suitable option for non-hostile situations. After items are rendered safe, the work party will proceed with the leak-sealing, packaging, and disposal operations.

Disposal

5-33. During combat operations chemical and biological ordnance can be disposed of if it will not increase the contamination of the area and will allow operations to continue immediately. When chemical or biological ordnance threatens operations or critical assets, the rendering safe of the ordnance may be required to allow asset recovery or mission continuation. Critical assets or areas that require immediate access may need additional EOD support.

5-34. Personnel conducting disposal operations must wear protective clothing that provides adequate protection against the agents. This protection is required only while actually working with the items. Set up the disposal area without protective clothing if the area is contamination free.

5-35. Before selecting the disposal site, consider the following:
- Direction of prevailing winds. This will allow populated areas to be warned and advised to don individual protective equipment; shelter in place or evacuate if they have no individual protective equipment.
- Elevation and openness of the terrain. An elevated and open terrain will allow the agent vapor clouds to disperse.
- Distance from any ammunition storage points, inhabited areas, training areas, highways, railroads, and airports. Consider fragmentation hazards as well as agent vapor cloud travel.
- Availability of an area of 60-meter radius cleared of combustibles. Label the area as a restricted area. Fence and post the area for regular usage with both visible and audible warning devices. Establish an exclusion area to prevent unprotected persons from exposure to agent vapors or clouds. Predetermine the size of the area and the amount of agent to be disposed of. Consider a 100 percent dissemination of agent and explosives when computing an exclusion area.
- Other features at the site, such as an aid station. EPDS, fire-fighting equipment, and a bunker or revetment for personnel protection from fragmentation. Also, communications between the disposal site and fire fighting and medical personnel is vital.
- Planners must be aware of the weather conditions and their factors that affect disposal operations.

5-36. Disposal operations are for emergency overseas operations only. In CONUS, contact 20th Support Command (CBRNE) for technical escort assistance needed for removal and disposal of chemical/biological materials. Federal and state regulations must be followed during recovery, transport, and disposition of biological and chemical agents. The Military Munitions Rule does not apply. All units should also obtain guidance from their appropriate service-specific arms control and treaty authority to ensure US treaty compliance is in accordance with the Chemical Weapons Convention.

EMERGENCY PERSONNEL DECONTAMINATION STATION

5-37. Personnel and equipment entering and exiting the incident site must take the route least likely to cause exposure or spread contamination. All personnel and equipment returning from a contaminated area must proceed through a decontamination station.

5-38. The EOD team must prepare to provide its own operational decontamination using the EPDS. The EOD unit commander may establish an EPDS for limited contamination control and decontamination. If CBRN decontamination support arrives, it sets up a decontamination station or relieves EOD on the site already established. The lack of a decontamination team or EPDS should not delay the EOD team's response.

5-39. The EPDS must be between the command post and the incident site and outside the fragmentation range of the explosive ordnance. It must be set up in a contamination free area clear of brush, trees, and other such vegetation. It must be upwind of the incident and at least 50 meters downwind from the command post.

5-40. A critical feature of the EPDS is the hot line (contamination control line). It is an imaginary line separating the contaminated area (hot zone) from the contamination reduction area (warm zone). It should be as close to the item as possible but outside its fragmentation radius. All personnel and equipment entering and leaving the incident area must process through the control point on the hot line. If the EOD commander or team leader considers it necessary, a shuffle pit may be established at the hot line.

5-41. The contamination control line (CCL) separates the contamination reduction area (warm zone) from the redress area (cold zone). Personnel do not cross into the redress area until they have been decontaminated. The contamination control line also prevents personnel from entering the contamination reduction area without wearing proper protective clothing. Decontaminate everything leaving the hot area and monitor for residue contamination before crossing the CCL.

5-42. The EPDS should be protected from the weather, if possible. It must be run by at least one EOD technician dressed in the proper protective clothing from the time personnel depart for the incident site until all personnel have been processed out.

5-43. The contamination reduction area may be contaminated by personnel returning from the incident area. Therefore, once decontamination operations begin, the contamination reduction area is considered contaminated. Make all efforts to control the spread of this contamination.

5-44. Upon completion of work at the site, decontaminate all exposed personnel. The extent and scope of the decontamination will depend upon the size and scope of the operation. The decontamination for a three-person team is much less elaborate than for an entire company. Follow the EPDS fundamentals (explained later) regardless of unit size.

5-45. How you arrange the EPDS will depend on the amount of contamination involved, the scope of the operation, and the terrain. Plan the EPDS to meet these requirements. Ensure personnel follow these four principles when using the EPDS:

- Move into the wind as undressing progresses.
- Decontaminate and remove the most heavily contaminated items of clothing first.
- Remove all articles of clothing worn at the incident site.
- Remove the mask and hood last before crossing into the cold zone.

5-46. The EOD Company should decontaminate as much of its own personnel protection items and mission-essential equipment as it can. Support personnel should decontaminate both the land in the area and nonessential equipment. When finished, EOD must mark the contaminated area of the EPDS for further decontamination.

5-47. The operation of the EPDS should be turned over to a CBRN decontamination platoon on its arrival. If EOD operations are completed before the CBRN decontamination platoon arrives, team members and other personnel should process through the EPDS. The hot line team must be prepared to turn over the operation to the incoming decontamination team. The contamination reduction area must be marked with contamination markers until the area is decontaminated. See figure 5-2 and 5-3 for example set ups for the EPDS.

Chapter 5

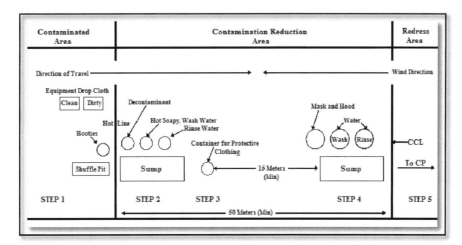

Figure 5-2. Example emergency personnel decontamination station (EPDS)

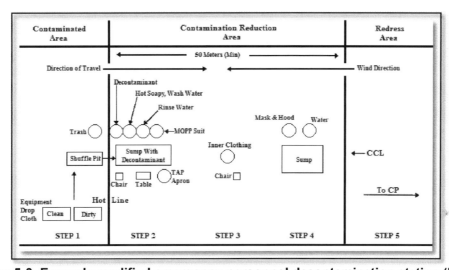

Figure 5-3. Example modified emergency personnel decontamination station (EPDS)

NUCLEAR ACCIDENT AND INCIDENT RESPONSE AND ASSISTANCE

5-48. Nuclear accident and incident response and assistance (NAIRA) is intended to minimize loss of life, personal injury, hazardous effects, and destruction of property and to secure nuclear material. Procedures described in this section are primarily intended for peacetime use. The DOD may provide support to the lead agency in an accident or incident involving radioactive materials. If required, EOD will respond from the closest military installation.

5-49. EOD forces identify and confirm the presence of nuclear weapons or materials; detect alpha, gamma, and beta hazards; perform actions to prevent a nuclear yield or high explosive detonation; perform initial packaging of materials; conduct emergency contamination control station operations for EOD personnel and their equipment only.

NUCLEAR WEAPONS

5-50. During operations in hostile or uncertain environments, it is possible that threat forces possess or employ nuclear weapons. These weapons may not have functioned as designed or may have been captured from the enemy in various conditions ranging from undamaged to extensively damaged. There is also the chance of a US or allied country nuclear weapon transportation accident in which US Army EOD may be the first on scene. Whatever the reason, the recovery of a nuclear weapon will require the involvement of EOD and supporting elements. Outlined below are some of the planning factors involved. Refer to AR 50-5 for responsibilities involving nuclear weapon incidents/accidents.

5-51. If the weapon is a threat system, Army EOD is responsible for preventing nuclear detonation or a high-explosive detonation. This includes conducting detection, identification, RSP, technical information gathering and preparing complete weapons or components for shipment.

5-52. If the weapon is a US system, the responsible service, either Navy or Air Force, will be notified. The specific service is ultimately responsible for any recovery actions required. Army EOD responsibility for the other service's weapons is to prevent a detonation or the spread of contamination. Once the other service's EOD personnel arrive, Army EOD personnel would help as needed and provide an EOD liaison between the services.

5-53. If it is a weapon system of an ally, Army EOD responsibility is to prevent detonation or the spread of contamination and to assist the friendly forces as authorized by command authorities.

5-54. The nuclear weapon recovery process could take several days and would require extensive logistical and operational support. The area commander must obtain the initial security and support elements. Some of the support necessary for the recovery operation includes: sustainment, engineer, RADCON and decontamination, transportation, security, aviation, and medical. Coordination with the other services and the host nation (if applicable) will also be needed to assist in the recovery operation.

OVERALL RESPONSIBILITIES

5-55. The headquarters having administrative control of the EOD company provides the location of the incident, identification of the explosive ordnance items, radiological hazard survey, and name and location of the individual authorized to grant access to the incident site. It must also identify security and evacuation measures being enforced and provide positive identification of the person authorized to receive the explosive ordnance items.

5-56. When the EOD company is assigned a nuclear accident/incident, it prepares to go to the scene by the fastest means. The team notifies the EOD battalion and coordinates with the NAIRA officer. If the EOD team leader is the first military representative on-site, he will assume control. The EOD commander (or senior EOD person present) is responsible for the safety of EOD personnel when they enter the hazard area.

5-57. Either the on-site commander or the NAIRA officer is responsible for the management of nuclear incident operations. In hostilities, the on-site commander may assume all responsibilities of a NAIRA officer. Resources available to the NAIRA officer for supporting operations include the following:

- Communications equipment (civilian and military)
- CBRN teams (survey and decontamination)
- Transportation (including air)
- Medical teams and facilities
- Engineer personnel and equipment
- EOD personnel
- Security forces (military and civilian)
- Military police

EOD OFFICER RESPONSIBILITIES

5-58. How the EOD officer handles the incident depends on resources available. It also depends on the current tactical situation, directives from the headquarters having operational control of the detachment, and quantity and type of weapons involved.

5-59. Specific responsibilities of the EOD officer include locating a potential command post site. The site may be approached from any direction if it is free of contamination. There is no need to monitor the route or command post area. Suspect areas must be approached from upwind. If contamination is suspected, the area must be checked with an alpha survey meter and a low-range gamma survey meter. The alpha survey meter may be used for spot checks, but the gamma survey meter must be used continuously. Surveys should begin at least one mile from the proposed command post site.

5-60. The EOD officer will:
- Designate the command post supervisor
- Designate a recorder
- Assign an equipment specialist
- Assign available personnel to work for the command post supervisor
- Designate the EOD team members (at least two EOD personnel trained in nuclear weapons)
- Designate the work party members. The work party must consist of at least one nuclear weapon trained EOD officer and another nuclear weapon trained Soldier (officer or enlisted)
- Assign an emergency contamination control station (ECCS) supervisor
- Identify the communications system to be used
- Determine respiratory protection required
- Ensure the EOD team and the work party is prepared to enter the incident site
- Determine and implement EMR hazard precautions as necessary. The NAIRA officer or on-scene commander must be notified of any EMR hazard and notify supporting elements

COMMAND POST SETUP AND RESPONSIBILITIES

5-61. The EOD team goes to the site as fast as possible. Once at the site, they must set up a command post. The EOD officer must consider several factors when setting up the command post:
- The area must be free of contamination and, if possible, have advantageous terrain features.
- The mode of transportation to be used to go to the incident site must be considered.
- If the item is unknown, an initial high explosive exclusion area with at least a 610-meter fragmentation radius plus a 50-meter ECCS distance must be established. The command post site must be upwind of the incident site or suspected contaminated area. It must also be close to, but separate from, support element operations. No other element should be established closer to the incident site than the EOD command post.

CP Supervisor

5-62. The command post supervisor organizes the command post for on-site operations and maintains contact with the headquarters having operational control of the EOD company. The command post supervisor is responsible for the following:
- Keeping the NAIRA officer informed of the incident progress.
- Inspecting the work party to ensure proper dress out and equipment setup.
- Ensuring dosimeters and film badges are worn properly. Dosimeters must be in the breast pocket of garments with pockets or on the upper arm of those without pockets. Film badges must be worn on the upper chest with the beta window uncovered and facing out.
- Adjusting the exclusion area as new information is received and making recommendations to the NAIRA officer or on-scene commander.
- Requesting disposition instructions for the contaminated waste, classified components, and other contaminated items.
- Notifying the decontamination team, of the type and amount of radiological contamination.
- Protecting classified information in the command post and making sure all classified materials are accounted for after the incident is completed.
- Preparing required reports and closing out the command post as required.
- Supervising the recorder, equipment specialist, and ECCS supervisor.
- Documenting radiation exposures using a radiation work permit IAW AR 40-13. Information from the radiation work party must be entered on the individual's DD Form 1141 (Record of Occupational Exposure to Ionizing Radiation) or automated dosimetry record. The radiation

work permit should fit the needs of the situation and be produced locally. The radiation work permit must include the following:
- Potential hazards.
- Protective clothing to be used.
- Special equipment requirements.
- Names and social security numbers of individuals working at the site.
- Serial numbers of personnel dosimeters.
- Film badge numbers.
- Time entering the site.
- Time departing the site.
- Initial dosimeter readings.
- Final dosimeter readings.

● Requesting and coordinating support needed for such things as protective measures, evacuation, medical, decontamination, and tactical security.

● Ensuring the RSP is completed.

Recorder

5-63. The recorder keeps a complete and accurate record of all actions and maintains communication with the work party. The recorder also relays information received to the command post supervisor and ensures that classified paperwork generated in the command post is properly marked and safeguarded.

Equipment Specialist

5-64. The equipment specialist is in charge of tool assembly, equipment checkout, respiratory apparatus setup, preoperational and operational checks of equipment, and charging pocket dosimeters.

5-65. The equipment specialist will also assist the work party in donning protective clothing and starting equipment. The equipment specialist also keeps a log of all equipment that is missing, broken, depleted, or destroyed. Finally, load the tools and equipment for return to the unit.

Emergency Contamination Control Station

5-66. Personnel and equipment entering and exiting the incident site must take the route least likely to cause exposure to or spread contamination. Therefore, all personnel and equipment returning from a contaminated area must proceed through a contamination control station. When there is no decontamination support, the EOD team may set up an ECCS. The EOD unit commander may establish an ECCS for limited contamination control and limited decontamination. When the nuclear emergency team arrives, it sets up a contamination control station and relieves EOD of the requirement.

5-67. The ECCS must be between the command post and the incident site and outside the fragmentation range of the munitions. It must be setup in an area free of contamination, upwind of the incident, and at least 50 meters downwind from the CP.

5-68. A critical feature of the ECCS is the hot line. It is an imaginary line separating the contaminated area from the contamination reduction area. The hot line should be as close to the item as possible but outside its fragmentation radius (610 meters for unknown nuclear weapons). All personnel and equipment entering and leaving the incident area must process through the control point on the hot line.

5-69. The CCL separates the contamination reduction area from the redress area. Personnel do not cross into the redress area until they are free of contamination or have acceptable levels of contamination. The CCL also prevents personnel from entering the contamination reduction area without wearing proper protective clothing.

5-70. The ECCS should be protected from the weather, if possible. It must be run by at least one EOD technician dressed in the proper protective clothing from the time personnel depart for the incident until all personnel have been processed out.

5-71. The contamination reduction area may be contaminated by personnel returning from the incident area. Therefore, once decontamination operations begin, the contamination reduction area is considered contaminated

5-72. The operation of the ECCS should be turned over to the CBRN decontamination platoon (the nuclear emergency team) on its arrival. If EOD operations are completed before the CBRN decontamination platoon arrives, team members and other personnel should process through the ECCS. The ECCS team should turn over the operation to the incoming decontamination platoon. The contamination reduction area must be marked with contamination markers until the area is decontaminated.

5-73. There is no need to process through an ECCS if no contamination was encountered. There is one exception: personnel who handle or come in contact with nuclear components must wash their hands and face using hot soapy water and rinse them with clear water. For additional information refer to Appendix D.

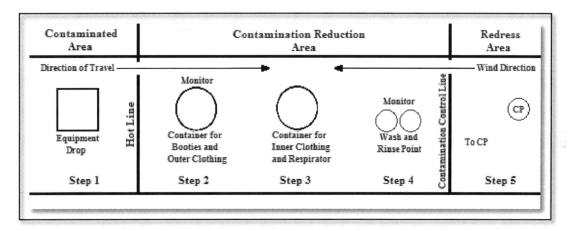

Figure 5-4. Emergency contamination control station (ECCS)

RESPOND TO A RADIOLOGICAL ATTACK

5-74. Radiological threats are often categorized with nuclear threats because the danger comes from the release of radiation; however, the major difference between the two is that the nuclear threat has a yield associated with it. The most likely form of a radiological attack would be from a radiological dispersal device, often called a "dirty bomb" or a radiological emitting device. While it is easier to acquire the components to construct a radiological dispersal device the result is much less damaging than a nuclear attack and the effects are less severe. The hazards associated with a nuclear attack include thermal radiation, ionizing radiation, blast and fallout. Due to the similarities to an improvised nuclear device the actions taken by the EOD company will be similar to responding to a nuclear device.

RESPOND TO DEPLETED URANIUM INCIDENTS

5-75. Accidents or incidents involving ammunition or armor containing depleted uranium (DU) pose special challenges. Technical Bulletin 9-1300-278 contains specific guidance for response to DU. EOD technician must be familiar with that guidance before responding to any potential DU incident.

EOD forces identify DU ordnance and equipment damaged by DU ordnance, detect alpha contamination, control the spread of contamination, package DU ordnance, and conduct emergency contamination control station operations for EOD personnel and their equipment only.

Appendix A
EOD Pre-Combat Checklist for Mounted Operations

This checklist is not intended to be an all inclusive list of tools and equipment for EOD operations. The EOD team leader is the deciding authority on what tools and equipment the EOD team will use based on the operation.

INDIVIDUAL
- ID Card, ID Tags, Battle Roster
- ACH & IBAA with ESAPI Plate, Knee & Elbow pads
- First Aid Pouch with Quick Clot
- Ammo (9mm, 5.56 drum, 7.62, 50 cal, 40mm)
- Night vision Serial #
- Water and MREs
- Safety Glasses and Goggles, Gloves
- Ear Plugs
- Seat Belt Cutter
- 3 Day Assault Pack
- Flash Lights with spare batteries
- ROE, EOF and BOLO list

MEDICAL
- CLS Bag
- Collapsible Litter
- 9 Line Report and Medevac Freq
- Medical Alert Tags
- Prior Heat and Cold Injured PAX ID'd

VEHICLE
- PMCS, Dispatch and license
- Vehicle Topped Off
- Batteries Chargers and Surge Protectors
- Spot Light, White Light, and IRs
- Escape Bottles
- Fire Extinguishers
- DVE
- GYROCAM/VOSS
- Load Plans Checked and Verified
- Slave Cables
- Commo and ECM check
- Basic Issue Items (Tow Bar or Tow Straps, Chains)
- GPS, Maps, Overlays, Compass, Binoculars
- Strip Maps, Convoy Brief
- 5 Gal Fuel Container, 5 Gal Water Container, Cooler with Ice

Appendix A

- VS 17 Panel, Smoke, IR Strobe, Chem Lights
- CROWS check and Test Fire

COMMUNICATIONS
- Antenna, Head Sets and Hand Mikes
- SOI, Call Signs and Freq's
- ANCD
- Long and Short Range Commo check
- Spare Batteries
- Hand Held Radios
- FBCB2, Flue Force Tracker Checked
- Way Points, NAI's, TAIs and Hot Spots Loaded
- Hub Batteries

CBRN
- Gas Mask
- JSLIST with Gloves and Boots
- M291 Kits
- M256
- M9 and M8 Paper
- ICAM, Draeger Kit and M-18 Kit
- UDR 13 and AN/PDR 77 or VDR 2
- Hasty Decon Kit
- NBC Marking Kit
- JCAD

SENSITIVE ITEMS

	M9	M4	M203	M249	M2	PEQ 2A	Optics
TL							
P2							
TM							

CLASS V
- M023, Charge Demo, C-4, 1 1/4 lb, M112
- M456, Cord Detonating
- M130, Cap Blasting, Electric M6
- M131, Cap Blasting, Non-Electric M7
- MN02, Cap Blasting, Non-Electric 500ft M12
- MN03, Cap Blasting, Non-Electric 1000ft M13
- MN08, Igniter, Fuse M81
- MN88, Minitube, 500Ft
- MN90, Minitube, 1000Ft
- M980, Charge, Deta Sheet, 36Ft Roll
- G900, Grenade, Hand Thermite, AN-M14
- A606, Cartridge, Caliber .50, API MK211
- Grenades (Smoke and Frag)
- Water Charges and Drop Charges Prepared

- PAN Ammo and 50 Cal. Carts

ROBOTS
- TALON OCU Serial#
- TALON Serial#
- Batteries Charged
- PackBot OCU Serial#
- PackBot Serial#
- Batteries Charged
- Long Range Antennas and Controllers

EOD TOOL KITS
- Bomb Suit, Helmet, Spare Batteries
- HAL Kit, Rope, Hot Stick, Painters Pole
- Remote Firing Device (Electric and NonEl), Firing Wire
- AN/PSS 14 and MIMID
- Evidence Recovery Kit (Camera, Packaging)
- Disruptors
- Laptop (AEODPS, JDIGS, CIDNE, ATTAC)
- Backpack ECM
- VSAT
- AHURA
- X-Ray
- MK 663, CMCs, WIF Kit, Velostat Bags
- Thermal
- SPEEDS Kit
- RECON Kit
- General Tool Kit and Pioneer Tools
- Cordless Tools
- Field Safe
- M107

CONVOY BRIEF
- Situation
  - Enemy Force
    - S-2 Update
    - Known or Suspected Locations of Combatants
    - Enemy COAs
    - Enemy TTPs in the AO
  - Friendly
    - Convoy make up
    - Adjacent unit coordination
    - Passage of line
- Mission
  - Who
  - What
  - Where
  - When
  - Why
- Execution

Appendix A

- Concept of Operation
    - Order of March
    - Routes
    - Density
    - Speed
    - Method of Movement
    - Defense on the Move
    - Start and Release Points, NAIs TAIs, Hot Spots
    - Control Measures
    - Actions at a Halt
    - Lighting
    - Fire Support
- Tasks to Maneuver Elements
- Tasks to Other Element
- Coordinating Instructions
    - Timeline
    - CCIR, PIR, EEIR, FFIR
    - Risk Reduction and Assessment
    - ROE and EOF
    - Environmental Considerations
    - Protection
- Service Support
    - Maintenance and Recovery
    - Medical
    - ROE and EOF
    - Personnel
        - EPW
        - Casualties
- Command and Signal
    - Convoy Commander
    - Mission Commander
    - Key Leaders
    - List Safe Havens Rally Points
    - SIO
    - CREW (Distribution and Compatibility)
- Rehearsals
    - Weapons Status
    - ROE and EOF
    - Medevac 9 Line
    - Call for Fire
    - Spot Report
    - Actions on Objective
    - CREW Check
- Drills
    - Rollover
    - Vehicle Fire
    - Vehicle Destroyed
- Contact
    - Near Ambush
    - Far Ambush

- Gunner Replacement
- React to an IED
- React to VBIED
- React to IDF
- Mine Strike
- Actions at a Halt
- Vehicle Recovery

Appendix B
EOD Pre-Combat Checklist for Dismounted Operations

This checklist is not intended to be an all inclusive list of tools and equipment for EOD operations. The EOD team leader is the deciding authority on what tools and equipment the EOD team will use based on the operation.

INDIVIDUAL
- ID Card, ID Tags, Battle Roster
- ACH & IBAA with ESAPI Plate, Knee & Elbow pads
- First Aid Pouch with Quick Clot
- Ammo (9mm, 5.56 drum, 7.62, 50 cal, 40mm)
- Night vision Serial #
- Water and MREs
- Safety Glasses and Goggles, Gloves
- Ear Plugs
- Seat Belt Cutter
- 3 Day Assault Pack
- Flash Lights with spare batteries
- ROE, EOF and BOLO list

MEDICAL
- CLS Bag
- 9 Line Report and Medevac Freq
- Medical Alert Tags
- Prior Heat and Cold Injured PAX ID'd

COMMUNICATIONS
- Antenna, Head Sets and Hand Mikes
- SOI, Call Signs and Freq's
- ANCD
- Long and Short Range Commo check
- Spare Batteries
- Hand Held Radios
- Way Points, NAI's, TAIs and Hot Spots Loaded
- Hub Batteries

CBRN
- Gas Mask
- JSLIST with Gloves and Boots
- M291 Kits
- M256
- M9 and M8 Paper
- UDR 13

Appendix B

- Hasty Decon Kit
- NBC Marking Kit

SENSITIVE ITEMS

	M9	M4	M203	M249	M2	PEQ 2A	Optics
TL							
P2							
TM							

CLASS V

- M023, Charge Demo, C-4, 1 1/4 lb, M112
- M456, Cord Detonating
- M130, Cap Blasting, Electric M6
- M131, Cap Blasting, Non-Electric M7
- MN02, Cap Blasting, Non-Electric 500ft M12
- MN03, Cap Blasting, Non-Electric 1000ft M13
- MN08, Igniter, Fuse M81
- M980, Charge, Deta Sheet, 36Ft Roll
- G900, Grenade, Hand Thermite, AN-M14
- A606, Cartridge, Caliber .50, API MK211
- Grenades (Smoke and Frag)
- Water Charges and Drop Charges Prepared
- PAN Ammo and 50 Cal. Carts

ROBOTS

- Lightweight Robot

EOD TOOL KITS

- Mini-HAL Kit, Rope, Hot Stick, Painters Pole
- Remote Firing Device (Electric and Non-el), Firing Wire
- Evidence Recovery Kit (Camera, Packaging)
- Disruptors
- Mine detector (not the Pss 14)

PATROL BRIEF

- Situation
  - Enemy Force
    - S-2 Update
    - Known or Suspected Locations of Combatants
    - Enemy COAs
    - Enemy TTPs in the AO
  - Friendly
    - Adjacent unit coordination
    - Passage of line
- Mission
  - Who
  - What
  - Where
  - When

- Why
- Execution
  - Concept of Operation
    - Order of March
    - Routes
    - Density
    - Method of Movement
    - Defense on the Move
    - Start and Release Points, NAIs TAIs, Hot Spots
    - Control Measures
    - Actions at a Halt
    - Lighting
    - Fire Support
  - Tasks to Maneuver Elements
  - Tasks to Other Element
  - Coordinating Instructions
    i. Timeline
    ii. CCIR, PIR, EEIR, FFIR
    iii. Risk Reduction and Assessment
    iv. ROE and EOF
    v. Environmental Considerations
    vi. Protection
- Service Support
  - Medical
  - ROE and EOF
  - Personnel
    o EPW
    o Casualties
- Command and Signal
  - Mission Commander
  - Key Leaders
  - List Safe Havens Rally Points
  - SIO
  - CREW (Distribution and Compatibility)
- Rehearsals
  - Weapons Status
  - ROE and EOF
  - Medevac 9 Line
  - Call for Fire
  - Spot Report
  - Actions on Objective
  - CREW Check
- Contact
  - Near Ambush
  - Far Ambush
  - Gunner Replacement
  - React to an IED
  - React to VBIED
  - React to IDF

Appendix C
Supporting Organizations

The complexity of current and emerging weapons systems require that EOD have comprehensive reach back capabilities up to the DOD and national levels. Additionally, EOD teams must have reach across and feedback capabilities to provide and receive near real time, relevant reporting on weapon systems, explosive ordnance and trends in order to populate databases with the most current data.

EXPLOSIVE ORDNANCE DISPOSAL DIRECTORATE

C-1. Develop, integrate, and synchronize Doctrine, Organization, Training, Materiel, Leadership and Education (DOTML) requirements for US Army EOD throughout Training and Doctrine Command; Coordinate Joint, Interagency, Intergovernmental, and Multinational EOD requirements in coordination with the Sustainment Center Of Excellence and DA G3/5/7. For more information go to (http://www.goordnance.army.mil).

G-38, ADAPTIVE COUNTER-IMPROVISED EXPLOSIVE DEVICE AND EXPLOSIVE ORDNANCE DISPOSAL SOLUTIONS DIVISION

C-2. The G-38 Division provides staff planning for assigned EOD strategic requirements in support of the geographic combatant commanders' operational plan. The G-38 Division accomplishes the planning by using the Joint Operation Planning and Execution System and coordinates the planning effort with the combatant commander's EOD staff officer.

C-3. G-38, within HQDA G3/5/7, manages the Army EOD Program in order to rapidly man, train, equip, and organize Army formations with the inherent ability to apply and defeat emergent asymmetric threats and adaptive networks including weapons of strategic of influence and asymmetric weapons, such as the IED, in support of unified land operations. For more information call (703-692-5975)

NAVAL EXPLOSIVE ORDNANCE DISPOSAL TECHNOLOGY DIVISION (NAVEODTECHDIV)

C-4. NAVEODTECHDIV exploits technology and intelligence to develop and deliver EOD information, tools, equipment, and life cycle support to meet the needs of joint service EOD operating forces and other specified customers. NAVEODTECHDIV manages the EOD database and provides a 24/7 reach back capability for UXO/IED/CBRN topics. For all information call (1-877-363-4636) / (1-877-363-2739) or email at EODTSC@navy.mil.

TERRORIST EXPLOSIVE DEVICE ANALYTICAL CENTER (TEDAC)

C-5. The mission of TEDAC is to prevent potential IED attacks by coordinating and managing the unified efforts of law enforcement, intelligence, and military assets to technically and forensically exploit all terrorist IEDs worldwide of interest to the US government. The information and intelligence derived from the exploitation of terrorist IEDs is used to provide actionable intelligence to anti-terror missions and to help protect the US military and multinational assets around the globe.

PICATINNY ARSENAL

C-6. Picatinny Arsenal is the Joint Center of Excellence for Armaments and Munitions, providing products and services to all branches of the US military. They specialize in the research, development,

Appendix C

acquisition and lifecycle management of advanced conventional weapon systems and advanced ammunition. Picatinny's portfolio comprises nearly 90 percent of the Army's lethality and all conventional ammunition for joint warfighters. For all EOD related information call (973-724-3868).

SANDIA NATIONAL LABORATORIES

C-7. Sandia National Laboratories is operated and managed by Sandia Corporation, a wholly owned subsidiary of Lockheed Martin Corporation. Sandia Corporation operates Sandia National Laboratories as a contractor for the US Department of Energy's National Nuclear Security Administration and supports numerous federal, state, and local government agencies, companies, and organizations. For more information contact the military liaison at (505-284-0537).

DEFENSE INTELLIGENCE AGENCY (DIA)

C-8. The Defense Intelligence Agency (DIA) is first in all-source defense intelligence to prevent strategic surprise and deliver a decision advantage to warfighters, defense planners, and policymakers. DIA deploys globally alongside warfighters and interagency partners to defend America's national security interests.

CBRNE ANALYTICAL & REMEDIATION ACTIVITY MOBILE EXPEDITIONARY LABORATORY

C-9. The CARA Mobile Expeditionary Laboratory deploys scientists to perform high-throughput chemical, explosives, and biological sample analysis to support Department of Defense combatant commanders, military installations, and support to US civil authorities if requested. CARA has three mobile lab packages: a light mobile expeditionary lab, a heavy mobile expeditionary lab, and a chemical air monitoring system, which deploy to support WMD elimination and remediation efforts in a forward deployed area. For more information call (410-436-6455).

UNITED STATES ARMY INTELLIGENCE AND SECURITY COMMAND (INSCOM)

C-10. INSCOM conducts a wide range of production activities, ranging from intelligence preparation of the battlefield to situation development, including signal intelligence analysis, imagery exploitation, and science and technology intelligence production. For more information call (703-428-4965).

NATIONAL GROUND INTELLIGENCE CENTER (NGIC)

C-11. NGIC produces and disseminates all-source integrated intelligence on foreign ground forces, systems, and supporting combat technologies to ensure that U.S. Forces have a decisive edge on any battlefield. NGIC supports US Army Forces during training, operational planning, deployment, and redeployment. NGIC maintains a counter IED targeting program CITP portal on the SECRET internet protocol router network web site that provides information concerning IED activities and incidents and NGIC IED assessments. In the IED fight, NGIC increases the capability of the multinational force to collect technical intelligence and provide dedicated intelligence fusion to support counter-insurgency operations. For more information call (434-980-7000).

CENTER FOR ARMY LESSONS LEARNED

C-12. The Center for Army Lessons Learned rapidly collects, analyzes, disseminates, and archives observations, insights, lessons learned, TTP and operational records in order to facilitate rapid adaptation initiatives and conduct focused knowledge sharing and transfer that informs the Army and enables operationally based decision making, integration, and innovation throughout the Army and within the JIIM environment. For more information call (913-684-2255/3035).

DEFENSE THREAT REDUCTION AGENCY (DTRA)

C-13. DTRA is the US Department of Defense's official Combat Support Agency for countering weapons of mass destruction. DTRA's programs include basic science research and development, operational support to US warfighters on the front line, and an in-house WMD think tank that aims to anticipate and mitigate future threats long before they have a chance to harm the US and our allies. For more information call (703-767-2000/2003).

ASYMMETRIC WARFARE GROUP (AWG)

C-14. The AWG conducts operations in support of joint and Army Forces commanders to mitigate and defeat specified asymmetric threats. The AWG assists in exploitation and analysis of asymmetric threats and provides advisory training for in-theater or pre-deployment forces.

TECHNICAL SUPPORT WORKING GROUP (TSWG)

C-15. The TSWG is the US national forum that identifies, prioritizes, and coordinates interagency and international research and development requirements for combating terrorism. The TSWG rapidly develops technologies and equipment to meet high priority needs of the combating terrorism community and addresses joint international operational requirements through cooperative research and development with major allies.

Appendix D
Contamination/Decontamination Station Setup

EPDS SETUP

STEP 1: EQUIPMENT DROP
- Equipment: Any material that prevents the contaminated equipment from contacting the ground, such as plastic bags, oilcloth, etc.
- Action: Place all equipment used at the incident site on the protective material. If a shuffle pit is used, all movement across the hot line is through it. Remove booties and put them in a container. Step across hot line onto the grate over the sump.

STEP 2: DECONTAMINATION
- Equipment: Containers (with sprayers if possible), for the following items: Decontaminant; hot soapy wash water; rinse water; decontaminant in the sump; and first aid for the agent(s) detected by IEP or WP.
- Action: Stand on grate over sump and spray, pour, or brush each person's impermeable protective clothing with decontaminant. Then spray, pour, brush, this time with hot soapy water, the individuals protective clothing. Finally spray, pour, or brush protective clothing with rinse water.

STEP 3: CLOTHING REMOVAL
- Equipment: Container for protective clothing.
- Action: Remove all clothing, except protective mask and hood, and place in container.

STEP 4: MASK AND HOOD REMOVAL AND SHOWER
- Equipment: Container, such as a plastic bag, for protective mask and hood; another container, such as a 5-gallon can, for wash water; grate for sump; and towels.
- Action: Step onto grate, take a deep breath, remove mask and hood and place in container. Then rinse head and upper body and resume breathing. Pour water over body and wash with soap, rinse body and proceed across contamination control line to redress area.

STEP 5: REDRESS AREA
- Equipment: Personnel Clothing, First-aid equipment and Self-sealing bags.
- Action: Redress and receive first aid (if required).

MODIFIED EPDS

STEP 1: EQUIPMENT DROP
- Equipment: Drop clothes, trash can lined with plastic bag, and shuffle pit (if needed).
- Action: Place all equipment used at the incident site on the protective material. If a shuffle pit is used, all movement across the hot line is through it. Place all trash in can provided. Step across the hot line onto the grate over the sump. Note: Hot line personnel should clean their hands with decontaminant and hot soapy water and rinse every time after touching a Soldier.

Appendix D

STEP 2: DECONTAMINATION

- Equipment: Five containers lined with plastic bags for toxicological agent protective apron, mission oriented protective posture (MOPP) Suit, decontaminant, hot soapy water, and rinse water. Each liquid-filled container should have a brush, sponge, and small can. Also needed is a chair, grate over a sump with decontaminant in it, a table with first aid for any agent encountered, M256 or M256A1 kit, grease pencil, scissors or knife, plastic bags, and drop cloths.
- Action: Stand on grate over sump and spray, pour, and brush protective boots, to include the bottom three or four inches of the MOPP pants with decontaminant, then hot soapy water, then rinse water. Wipe mask, hood, apron, and gloves with a damp sponge, first with decontaminant, then with hot soapy water, and then with rinse water. Remove apron and place it in apron bucket. Roll the hood: leave the zipper closed and lift the hood straight up off the shoulders by grasping the straps. Pull the hood over the head until most of the back of the head is exposed. Do not pull the hood completely over the face. Then roll the hood, starting at the chin and working around the mask. Roll it tightly, but do not pull it completely off the back of the head. You may place your hand over the voicemitter to prevent the mask seal from being broken. Remove the overgarment as described below.
- Jacket: Untie the cord and unfasten the snaps on the front of the jacket. Unzip the jacket and unsnap the snaps in the back of the jacket from the trousers. Pull the jacket off, one arm at a time, turning the jacket inside out. Make a fist as each sleeve is pulled off to prevent the gloves from coming off. Place the jacket in the bucket marked MOPP.
- Trousers: Remove the trousers by first opening the trouser cuffs, then the waist snap, zipper, and, if necessary, the waist tabs. Grasp the trousers by the cuff while the soldier pulls one leg at a time from the trousers. Place the trousers in the bucket marked MOPP.
- Overboots: Have the soldier sit in a chair at the end of the sump. Untie or cut the strings of the Soldier's overboots. Pull them off, one leg at a time (right boot first). As the overboots are removed, the Soldier steps on the ground at the end of the sump. Place the overboots in the bucket marked MOPP.
- Rubber Gloves: The Soldier removes his or her rubber gloves. Hot line personnel will assist so that the Soldier does not touch the outside of the gloves. Place the gloves in the bucket marked MOPP. Proceed to inner clothing removal.

STEP 3: INNER CLOTHING REMOVAL

- Equipment: Chair, container for inner clothing (lined with plastic bag), and shower shoes.
- Action: Remove all clothing, except protective mask and hood, and place it in the container for inner clothing. Proceed to the shower point.

STEP 4: MASK AND HOOD REMOVAL AND SHOWER

- Equipment: Two containers lined with plastic bags; one for mask and hood and the other for wash water. A table with towels and soap.
- Action: Step onto the grate, take a deep breath, remove mask and hood, and place them in the container. Then rinse head and upper body and resume breathing. Pour water over body and wash with soap. Rinse body and proceed across contamination control line to the redress area.

STEP 5: REDRESS AREA

- Equipment: Redress kit, first aid equipment, and litter.
- Action: Redress and receive first aid if required.

ECCS SETUP

STEP 1: EQUIPMENT DROP AND BOOTIE REMOVAL
- Equipment: Any material, such as plastic bags or oilcloth, which prevents the contaminated equipment from contacting the ground.
- Action: Place equipment on material provided. Remove film badge and dosimeters. Read dosimeters. Remove booties and step over the hot line. Place booties in container.

STEP 2: OUTER CLOTHING REMOVAL AND MONITORING
- Equipment: Containers, such as plastic bags, for booties and outer clothing. Radiacmeters for contamination encountered. Dosimeter register.
- Action: Remove outer clothing and gloves, taking care not to touch outer clothing. Step forward. ECCS personnel monitor for contamination. If no contamination is detected or contamination has been determined to be acceptable, the person removes respirator, washes, and moves to redress area. If any contamination is detected above permissible levels, the person immediately goes to Step 3.

STEP 3: INNER CLOTHING AND RESPIRATOR REMOVAL
- Equipment: Container, such as plastic bag, for inner clothing and respirator.
- Action: Remove inner clothing, including inner gloves and respirator, and place it in the container.

STEP 4: WASH AND RINSE POINT
- Equipment: Container, such as plastic tubs, for hot soapy water and rinse water. Also needed are towels, radiacmeters, cornmeal, and powdered soap.
- Action: Wash body with soapy water, paying particular attention to the fingernails and hairy portions of the body. Then rinse the body with clean water, dry, and monitor body for contamination. If contamination above permissible levels is detected, repeat the washing, rinsing, and drying process, then remonitor. If the second washing doesn't work, a mixture of 50 percent powdered soap and 50 percent cornmeal mixed with water is massaged onto the contaminated area for five minutes. The mixture is rinsed from the body and the person is remonitored. If contamination is still detected, medical assistance should be requested. If there is no contamination or contamination is within acceptable limits, the person crosses the contamination control line and redresses after washing face and hands.

STEP 5: REDRESS AREA
- Equipment: Personal Clothing, First-aid equipment, Self-sealing bags.
- Action: Redress and receive first-aid (if required).

Glossary

Terms for which ATP 4-32 is the proponent are marked with an asterisk (*). The proponent publication for other terms is listed in parentheses after the definition.

SECTION I – ACRONYMS AND ABBREVIATIONS

ADP	Army Doctrinal Publication
ADRP	Army Doctrinal Reference Publication
ARNG	Army National Guard
AO	area of operations
AR	Army Regulation
ARSOF	Army Special Operations Forces
ATP	Army Techniques Publication
ATTP	Army Tactics Techniques and Procedures
AWG	Assymetric Warfare Group
BCT	brigade combat team
CAIRA	Chemical Accident or Incident Response and Assistance
CBRN	chemical, biological, radiological, and nuclear
CFR	Code of Federal Regulations
CJTF	combined joint task force
CCL	contamination control line
CONUS	Continental United States
CREW	counter radio controlled IED electronic warfare
DA	Department of the Army
DIA	Defense Intelligence Agency
DNA	deoxyribonucleic acid
DOD	Department of Defense
DOS	Department of State
DSCA	Defense Support of Civil Authorities
DTRA	Defense Threat Reduction Agency
DU	depleted uranium
ECCS	emergency contamination control station
ECM	electronic countermeasure
EMR	electromagnetic radiation
EOD	explosive ordnance disposal
EODIMS	Explosive Ordnance Disposal Information Management System
EPDS	emergency personnel decontamination station
FBI	Federal Bureau of Investigation
FM	Field Manual
HMA	Humanitarian Mine Action
IED	improvised explosive device
INSCOM	Intelligence and Security Command

JP	Joint Publication
METT-TC	mission, enemy, terrain and weather, troops and support available, time available, civil considerations
MOA	memorandum of agreement
MOPP	mission oriented protective posture
MPCV	mine protected clearance vehicle
MTTP	Multi-Service Tactics, Techniques, and Procedures
NAIRA	nuclear accident and incident response and assistance
NBC	nuclear, biological, and chemical
NCO	noncommissioned officer
NGIC	National Ground Intelligence Center
ODA	Operational Detachment Alpha
ODB	Operational Detachment Bravo
OE	operational environment
OPCON	operational control
PPD	Presidential Policy Directive
RCIED	Radio Controlled Improvised Explosive Device
RSP	render safe procedures
SOF	Special Operations Forces
SOP	standard operating procedures
STANAG	Standardization Agreement
TEDAC	Terrorist Explosive Device Analytical Center
TSWG	Technical Support Working Group
TTP	tactics, techniques and procedures
USSS	United States Secret Service
UXO	unexploded explosive ordnance
WMD	weapons of mass destruction

SECTION II – TERMS

electromagnetic radiation

Radiation made up of oscillating electric and magnetic fields and propagated with the speed of light. (JP 6-01)

electronic warfare

Military action involving the use of electromagnetic and directed energy to control the electromagnetic spectrum or to attack the enemy. (JP 3-13.1)

exploitation

Taking full advantage of any information that has come to hand for tactical, operational, or strategic purposes. (JP 2-01.3)

explosive ordnance

All munitions containing explosives, nuclear fission or fusion materials, and biological and chemical agents. (JP 3-34)

explosive ordnance disposal

The detection, identification, on-site evaluation, rendering safe, recovery, and final disposal of unexploded explosive ordnance. Also called **EOD.** (JP 3-34)

explosive ordnance disposal unit

Personnel with special training and equipment who render explosive ordnance safe, make intelligence reports on such ordnance, and supervise the safe removal thereof. (JP 3-34)

explosive ordnance disposal incident

The suspected or detected presence of unexploded explosive ordnance, or damaged explosive ordnance, which constitutes a hazard to operations, installations, personnel or material. Not included in this definition are the accidental arming or other conditions that develop during the manufacture of high explosive material, technical service assembly operations or the laying of mines and demolition charges. (AAP 6)

explosive ordnance disposal procedures

Those particular courses or modes of action taken by explosive ordnance disposal personnel for access to, diagnosis, rendering safe, recovery and final disposal of explosive ordnance or any hazardous material associated with an explosive ordnance disposal incident. **a. Access procedures** – Those actions taken to locate exactly and to gain access to unexploded explosive ordnance.**b. Diagnostic procedures** – Those actions taken to identify and evaluate unexploded explosive ordnance.**c. Render-safe procedures** – The portion of the explosive ordnance disposal procedures involving the application of special explosive ordnance disposal methods and tools to provide for the interruption of functions or separation of essential components of unexploded explosive ordnance to prevent an unacceptable detonation. **d. Recovery procedures** – Those actions taken to recover unexploded explosive ordnance.**e. Final disposal procedures** – The final disposal of explosive ordnance which may include demolition or burning in place, removal to a disposal area or other appropriate means. (AAP 6)

homeland defense

The protection of United States sovereignty, territory, domestic population, and critical defense infrastructure against external threats and aggression or other threats as directed by the President. (JP 3-27)

homemade explosive

Non-standard explosive mixtures / compounds which have been formulated / synthesized from available ingredients. Most often utilized in the absence of commercial / military explosives. (WTI IED Lexicon 4th edition)

humanitarian mine action

Activities that strive to reduce the social, economic, and environmental impact of land mines, unexploded ordnance and small arms ammunition - also characterized as explosive remnants of war. (JP 3-15)

improvised explosive device

A weapon that is fabricated or emplaced in an unconventional manner incorporating destructive, lethal, noxious, pyrotechnic, or incendiary chemicals designed to kill, destroy, incapacitate, harass, deny mobility, or distract. Also called **IED**. (JP 3-15.1)

unexploded exlposive ordnance

Explosive ordnance which has been primed, fused, armed or otherwise prepared for action, and which has been fired, dropped, launched, projected, or placed in such a manner as to constitute a hazard to operations, installations, personnel, or material and remains unexploded either by malfunction or design or for any other cause. Also called **UXO**. (JP 3-15)

weapons of mass destruction

Chemical, biological, radiological, or nuclear weapons capable of a high order of destruction or causing mass casualties and exclude the means of transporting or propelling the weapon where such means is a separable and divisible part from the weapon. (JP 3-40)

weapons technical intelligence

A category of intelligence and processes derived from the technical and forensic collection and exploitation of improvised explosive devices, associated components, improvised weapons, and other weapon systems. (JP 3-15.1)

References

REQUIRED PUBLICATIONS
These documents must be available to intended users of this publication.

ADRP 1-02, *Operational Terms and Military Symbols*, 31 August 2012

JP 1-02, *Department of Defense Dictionary of Military and Associated Terms*, 8 November 2011

RELATED PUBLICATIONS
These documents contain relevant supplemental information.

ARMY PUBLICATIONS

ADP 3-0, *Unified Land Operation*, 10 October 2011.

ADP 3-28, *Defense Support of Civil Authorities*, 26 July 2012.

ADRP 3-0, *Unified Land Operations*, 16 May 2012.

ADRP 3-05, *Special Operations*, 31 August 2012.

ADRP 3-37, *Protection*, 31 August 2012.

AR 40-13, *Radiological Advisory Medical Teams*, 1 October 2012.

AR 50-5, *Nuclear Surety*, 1 August 2000.

AR 50-6, *Chemical Surety*, 28 July 2008.

AR 75-14, *Interservice Responsibilities for Explosive Ordnance Disposal*, 14 February 1992.

AR 75-15, *Policy for Explosive Ordnance Disposal*, 22 February 2005.

AR 190-11, *Physical Security of Arms, Ammunition, and Explosives*, 5 September 2013.

AR 190-45, *Law Enforcement Reporting*, 30 March 2007.

AR 525-27, *Army Emergency Management Program*, 13 March 2009.

ATTP 4-32.16, *EOD Multi-Service Tactics, Techniques, and Procedures for Explosive Ordnance Disposal*, 20 September 2011.

ATTP 4-32.2, *MTTP for Unexploded Ordnance*, 20 September 2011.

ATTP 3-90.15, *Site Exploitation Operations*, 8 July 2010.

FM 3-11.4, *Multiservice Tactics, Techniques, and Procedures for Nuclear, Biological, and Chemical (NBC) Protection*, 2 June 2003.

FM 27-10, *The Law of Land Warfare*, 18 July 1956.

TC 3-90.119, *US Army Improvised Explosive Device Defeat Training*, 23 July 2009.

TB 9-1300-278, *Guidelines for Safe Response to Handling, Storage, and Transportation Accidents Involving Army Tank Munitions or Armor Which Contain Depleted Uranium*, 23 February 2001.

JOINT PUBLICATIONS

CJCSI 3207.1, *Military Support to Humanitarian Demining Operations*, 1 November 2008.

JP 1-02, *DOD Dictionary of Military and Associated Terms*, 15 July 2012.

JP 3-0, *Joint Operations*, 11 August 2011.

JP 3-05, *Special Operations*, 18 April 2011.

JP 5-0, *Joint Operation Planning*, 11 August 2011.

JP 3-15.1, *Counter IED Operations*, 9 January 2012.

ALLIED PUBLICATIONS

STANAG 2143, *Explosive Ordnance Reconnaissance/Explosive Ordnance Disposal*, 16 September 2005.

STANAG 2389, *Minimum Standards of Proficiency for Trained Explosive Ordnance Disposal Personnel*, 6 January 2009.

NATP Allied Technical Publication 72, *Inter-Service Explosive Ordnance Disposal Operations on Multinational Deployments*, 16 May 2006.

DEPARTMENT OF DEFENSE PUBLICATIONS

DOD Directive 3025.13, *Employment of DOD Capabilities in Support of the U.S. Secret Service (USSS), Department of Homeland Security (DHS)*, 8 October 2010.

DOD Directive 3025.18, *Defense Support of Civil Authorities (DSCA)*, 29 December 2010.

DOD 3150.8-M, DOD *Response to Nuclear and Radiological Incidents*, 20 January 2010.

DOD 5100.76-M, *Physical Security of Sensitive Conventional Arms, Ammunition, and Explosives*, 17 April 2012.

DOD Instruction 3025.21, *Defense Support of Civilian Law Enforcement Agencies*, 27 February 2013.

DOD Instruction 4650.01, *Policy and Procedures for Management and Use of the Electromagnetic Spectrum*, 9 January 2009.

Sustaining US Global Leadership: Priorities for 21st Century Defense, 5 Jan 2012.

Strategy for Homeland Defense and Defense Support of Civil Authorities, 21 July 1996.

PRESIDENTIAL DIRECTIVES

PPD 1, *Organization of the National Security Council*, 13 February 2009.

PPD 2, *Implementation of the National Security Strategy for Countering Biological Threats*, 23 November 2009.

PPD 7, *Critical Infrastructure Identification, Prioritization, and Protection*, 17 December 2003.

PPD 8, *National Preparedness*, 20 March 2011.

PPD 17, Countering Improvised Explosive Devices, 14 June 2012.

CODE OF FEDERAL REGULATION

Title 10, *Armed Forces.*

Title 29, *Labor, Part 1200 and 1910.120.*

Title 32, National Guard.

Title 40, *Protection of Environment, Part 260-265.*

Title 49, *Transportation.*

TOE

TOE 09627R000 HHD OD Group (EOD).

TOE 09446R000 HHD OD Battalion (EOD).

TOE 09440R000 OD Company, EOD (RECAP).

TOE 09747R100 OD Company (EOD) (CONUS SUPT).

TOE 09747R000 EOD WMD Company.

TOE 09547RB00 EOD Platoon.

REFERENCED FORMS

Unless otherwise indicated, DA forms are available on the APD Web site (http://www.apd.army.mil).

DA Form 2028, *Recommended Changes to Publications and Blank Forms.*

DA Form 3265, *Explosive Ordnance Incident Report.*

DA Form 4137, *Evidence/Property Custody Document.*

DD Form 1141, *Record of Occupational Exposure to Ionizing Radiation.* (Available thru normal supply channels).

WEBSITES

Naval EOD Technology Division, <http://www.navsea.navy.mil/nswc/eodtechdiv/default.aspx>

Department of Homeland Security < http://www.dhs.gov/>

United States Environment Protection Agency <http://www.epa.gov/

Index

0
0,5s and 25s, 2-17

A
adapt the force, 2-4
air assault, 2-12
airfield clearance, 2-12
airfield seizure, 2-8
ammunition supply point, 1-11
Army National Guard, 1-7
attack the network, 2-2, 2-4

B
blast, 2-22
blow in place, 2-5
booby trap, 2-22
brigade combat team, 1-8

C
capabilities, 2-5, 3-4
captured enemy ammunition, 1-12
captured enemy ammunition/munitions, 1-11
captured material exploitation, 1-12
chemical, biological, radiological, and nuclear, A-2, B-1
chemical, biological, radiological, nuclear, 5-1
chemical, biological, radiological, nuclear, 2-5
combatant commander, C-1, C-2
combined arms maneuver, 1-1
combined joint task force, 1-13, 2-3
communication, 2-6
communication architecture, 2-6
contingency, 2-1
cordon and search, 2-12
corps, 1-8
counter radio controlled IED electronic warfare, 2-2, 2-5, 2-7
counter-IED, 1-2, 1-8, 1-9, 1-13, 2-2, 2-5

counter-IED operations, 2-4

D
defeat the device, 2-2, 2-4
Department of Justice, 3-8
Department of State, 1-9, 1-12, 3-8
diplomatic security service, 1-13
direct support, 2-5
dismounted CREW, 2-7
dismounted patrol, 2-13
disposal, 1-11, 2-5, 3-4
dispose, 2-22
division, 1-8

E
electronic countermeasure, 3-2, A-1
emergency response, 2-15
emergency response operations, 2-9
EOD 9 Line, 2-9, 2-15, 2-16, 2-27
EOD battalion, 1-2, 1-4, 1-5, 1-8, 1-10, 1-11, 1-13, 2-2, 2-3
EOD battalion force structure, 1-4
EOD categories, 2-10
EOD company, 1-5, 1-11, 2-2, 2-3, 2-5
EOD company (CONUS Support), 1-7
EOD company WMD, 1-6
EOD company force structure, 1-5
EOD company WMD force structure, 1-6
EOD group, 1-2, 1-5, 1-8, 1-10, 1-11, 1-13, 2-2, 2-3
EOD group force structure, 1-3
EOD information management system, 3-6
EOD platoon, 1-2, 1-5, 1-6, 1-10, 1-11, 2-5
EOD platoon force structure, 1-6
EOD staff non-commissioned officer, 1-8
EOD staff officer, 1-8

EOD team, 2-5, 3-5
EOD team operations, 2-11
exploitation, 1-2, 1-10, 2-2, 2-5
explosive ordnance, 2-6, 2-16, 2-22
explosive precursor, 2-22

F
follow on call, 2-27
foreign internal defense, 1-10
foreign ordnance, 1-12
fragmentation, 2-22

G
geographic combatant commander, 1-9
Guardian Angel, 2-15

H
homemade explosive, 1-12
homemade explosives, 2-22
humanitarian mine action, 1-9, 1-12

I
improvised explosive device, 2-4, 2-5, 2-21
incident, 2-9
incident reporting, 2-9
intelligence, 2-7, 3-2

J
joint, 1-2, 1-6
joint operational phases, 1-8
joint operations, 2-3
Joint Prisoner of War/Missing in Action Accounting Command, 1-10

L
logistical support, 2-10

M
maneuver, 2-5, A-4, B-3
medical, 2-11
military to military engagement
misfired munitions, 1-11
mission command, 1-2, 1-13, 2-2, 2-3, 2-5
mission statement, 1-1
mission variables, 2-1

Index

mounted CREW, 2-7
movement, 2-16
multinational, 1-2
multinational operations, 2-3

O

operational environment, 2-1

P

Phase 0, 1-9
Phase I, 1-10
Phase II, 1-11
Phase III, 1-11
Phase IV, 1-12
Phase V, 1-13
planned operaions, 2-8
planned operation, 2-7
planned operations, 2-11
planning considerations, 2-7
Posse Comitatis Act, 3-2
post blast analysis, 1-10, 2-5, 2-7, 2-23
protection principles, 1-1

Q

questioning techniques, 2-18

R

render safe, 1-11, 2-5
render safe procedure, 2-22
render safe procedures, 2-5
response team, 1-6
route clearance, 2-11
rules of allocation, 1-3, 1-4, 1-5

S

scene turnover, 2-27
search, 2-25
security, 2-5, 2-10, 2-19, 2-27, 2
security force assitance, 1-10
small rewards program, 2-14
special operations force, 1-10, 1-12, 2-5
special operations forces, 4-1
special programs, 1-12, 2-5
stuck round, 1-11

T

tactical post blast, 2-23
targeting, 1-2
Team Leader certification, 1-7
technical intelligence, 1-2, 1-10, 2-5
tools, 2-16
training and partnership, 2-15

U

unexploded explosive ordnance, 1-12, 2-5, 2-22
US Secret Service, 3-8

V

Very Important Persons Protection Support Activity, 3-3

W

weapons buyback, 2-14
weapons of mass desctruction elimination, 1-13
weapons of mass destruction, 1-11, 2-5
wide area security, 1-1

Made in the USA
Columbia, SC
23 March 2019